STRATHCLYDE UNIVERSITY LIBRARY

ML

D1485675

ANDERSONIAN LIBRARY

★

WITHDRAWN
FROM
LIBRARY
STOCK

★

UNIVERSITY OF STRATHCLYDE

RYE:

Production, Chemistry, and Technology

Edited by Dr. Walter Bushuk Department of Plant Science,
University of Manitoba
Winnipeg, Manitoba, Canada

Published by the
**American Association of Cereal Chemists, Incorporated
St. Paul, Minnesota**

Copyright © 1976 by the
AMERICAN ASSOCIATION OF CEREAL CHEMISTS, INC.
St. Paul, Minnesota

*All rights reserved. No part of this
book may be reproduced in any
form by photostat, microfilm,
retrieval system, or any other
means, without written permission
from the publishers.*

LIBRARY OF CONGRESS CATALOG CARD NUMBER: 76-029382

ISBN 0-913250-11-2

Printed in the UNITED STATES OF AMERICA

D
633.14
RYE

PREFACE

The awareness of the need for an all-inclusive publication on rye arose when I could not find the information needed to answer an inquiry regarding milling and baking quality assessment of this grain. A proposal that the AACC should publish a monograph on rye was made to the Committee on Publications through its Chairman, Professor Paul Mattern, Department of Agronomy, University of Nebraska. The proposal was unanimously approved.

It is intended that the book should be useful to the student and scientist interested in any aspect of rye. Accordingly, it covers production, breeding, properties, and processing. For the content of this book, the credit must go to the authors of the individual chapters. The authors have given much of their time and expert knowledge in the preparation of the manuscripts. It is my view that their wide range of experience adds significantly to the value of the book.

In editing the monograph, I have attempted to achieve some uniformity among chapters and to eliminate obvious duplication without completely destroying the individuality of the authors with widely varying backgrounds and interests. I commend their excellent cooperation and willingness to take their manuscripts through several revisions.

The editorial assistance of my colleagues Dr. C. C. Bernier, Dr. F. W. Hougen, Dr. J. P. Gustafson, Dr. L. J. LaCroix and Mr. I. Levi is gratefully acknowledged. I sincerely acknowledge the excellent support and cooperation that I received from Mr. R. J. Tarleton, Executive Vice President, AACC, and Ms. J. F. Sorensen, the technical editor of this monograph. Finally, my sincere thanks go to Mrs. Sylvia Kusmider for her typing services.

Walter Bushuk

AMERICAN ASSOCIATION OF CEREAL CHEMISTS, INC.

Volume V
Monograph Series

CONTRIBUTORS

W. Bushuk, Department of Plant Science, The University of Manitoba, Winnipeg, Manitoba, Canada

W. P. Campbell, C.S.I.R.O. Wheat Research Unit, North Ryde, Australia

E. Drews, Federal Research Institute of Cereal and Potatoe Processing, Detmold, Federal Republic of Germany

L. E. Evans, Department of Plant Science, The University of Manitoba, Winnipeg, Manitoba, Canada

T. A. Rozsa, 353 W. Broadway, Winona, Minnesota

G. J. Scoles, Department of Plant Science, The University of Manitoba, Winnipeg, Manitoba, Canada

W. Seibel, Federal Research Institute of Cereal and Potatoe Processing, Detmold, Federal Republic of Germany

D. H. Simmonds, C.S.I.R.O. Wheat Research Unit, North Ryde, Australia

Stanislaw Starzycki, Plant Breeding and Acclimatization Institute, Radzików, Poland

CONTENTS

RYE:

Production, Chemistry, and Technology

CHAPTER 1

HISTORY, WORLD DISTRIBUTION, PRODUCTION, AND MARKETING

W. BUSHUK
Department of Plant Science
The University of Manitoba, Winnipeg, Manitoba, Canada

I. INTRODUCTION

Rye (*Secale cereale* L.) is second only to wheat as the grain used most commonly for the production of bread. Because of the extreme hardiness of the rye plant and its ability to grow in sandy soils of low fertility, rye can be grown in areas that are generally not suitable for growing other cereal grains. Greatest production is in the cool temperate zones of the world, but it can also grow in the semiarid regions near deserts and at high altitudes. It enjoys the widest distribution of all the cereal crops.

II. ORIGIN AND HISTORY

The primary center of origin of rye appears to be southwestern Asia, essentially the same as the area of origin of wheat, oats, and barley (Deodikar, 1963). Rye is not as old as wheat. There is no trace of cultivated rye in ancient Egyptian monuments nor is it mentioned in any of the ancient writings. It is mentioned in early northern European writings which suggests that it was first cultivated in this area. Grains found in the Neolithic sites of Austria and Poland are considered to be of "wild" origin.

Rye moved from its center of origin to northern Europe sometime during the first millenium B.C. The exact route of this migration is not known. One possible route is from Asia Minor northwards into Russia and then westwards into Poland and Germany (Scheibe, 1935; Kuckuck, 1937). According to Popov (1939), a second possible route of migration is from Turkey into Europe across the Balkan Peninsula. From there, northern European rye gradually spread throughout most of Europe and was eventually brought to North America and western South America with the settling of these areas by Europeans in the 16th and 17th centuries. During this period, it gradually spread across the southern

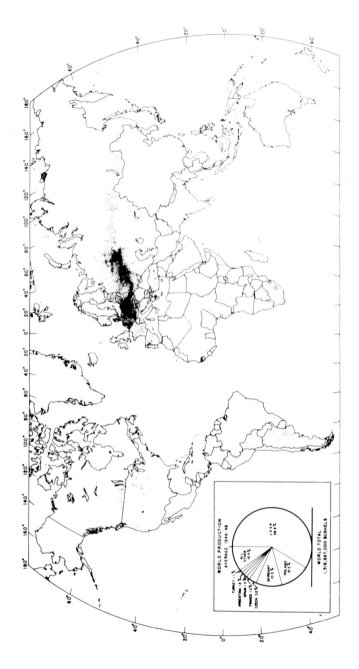

Figure 1. World rye distribution. Each dot represents one million bushels, 1946-48 average. (Leonard and Martin, 1963).

fringe of U.S.S.R. into Siberia. In the 19th and 20th centuries it was introduced to Argentina, southern Brazil, Uruguay, Australia, and South Africa. Figure 1 shows the areas in the world where rye is grown.

III. CLASSIFICATION

Rye is a member of the grass family and is subclassified into the genus *Secale*. Only one species of rye, *Secale* (S.) *cereale* L., is extensively cultivated. According to Carleton (1919), Hackel divided rye into two species, *Secale* (S.) *fragile* Bieberst and *S. cereale* L. *S. fragile* is grown to a limited extent in southwestern Asia. The wild progenitor of *S. cereale* has not been definitely identified although some taxonomists believed that it evolved from *Secale montanum* Gus., a perennial that grows wild in southern Europe and central Asia.

Most of the cultivated rye is of the type with seven pairs of somatic chromosomes. An artificially produced tetraploid type with 14 pairs of chromosomes is grown to a limited extent in Europe.

The number of different varieties of rye grown around the world is relatively low, especially when compared with wheat. Considerably less effort has been expended in the improvement and development of rye varieties than of most other cereal grains. Because rye is a cross-pollinated crop, it is extremely difficult to keep rye varieties pure.

Most of the rye is grown as a fall-sown annual. This is generally called winter rye. Because of its superior winter hardiness, winter rye can be grown successfully in areas where the climate is too severe for winter wheat. Some spring rye is grown in areas where the winters are too severe for winter rye production. The spring varieties are generally inferior in agronomic and end use qualities.

IV. AREA, YIELD, AND QUANTITY OF WORLD PRODUCTION

The area of cultivated land of the world devoted to the growing of rye (FAO, 1972) has decreased substantially over the past decade (Figure 2). In 1961, 28.5 million hectares were harvested; by 1972 this figure had decreased to 17.5 million hectares, a drop of 35%. During the same period, total production decreased from 35 million to about 31 million metric tons, a drop of only about 11%. The substantial decrease in area was largely offset by the marked increase in yield. Yields in the early 1960's were as low as 11.5 centals per hectare. After 1964, yields increased continuously, reaching the highest recorded figure of almost 17 centals per hectare in 1971—an increase of about 48%. This substantial increase was achieved through improvement of agronomic practices, especially in the use of chemical fertilizers and crop rotation, and through improvement of varieties and elimination of the use of low-fertility land.

Relative to other major cereal crops, rye was in eighth position over the past decade (Table I). Its production was about one-tenth that of the three major crops—wheat, rice, and maize.

Rye is a particularly important crop in the U.S.S.R., Poland, and Germany (Table II). Approximately one-third of the world production was in the U.S.S.R. in 1972. A slightly lower proportion was produced in Poland where rye is the leading cereal crop; it exceeded wheat by nearly 60%. In the German Federal Republic rye formed as much as 31% of the combined rye and wheat production. The equivalent figure for the German Democratic Republic was 41%; and for Austria, where the production of rye and wheat is quite small, the figure was 32%. Rye occupies an important economic position in many other countries.

In spite of world food shortages, it is not envisaged that rye production will increase in the future. Because of the hardiness of the plant, however, interest in the crop will probably remain high, and it should continue to be an important crop.

V. USES, MERITS, AND DEFICIENCIES

The rye crop has many uses. It is used as flour for bread, as grain for livestock feed, and as a green plant for livestock pasture.

Of the cereal grains, only wheat and rye produce flours that can be used for the production of leavened bread. Rye is inferior to wheat in breadmaking quality; its dough lacks elasticity and gas retention properties. Rye flour can be used alone to produce the so-called black bread used extensively in eastern Europe and parts of Asia. In many countries, a lighter rye loaf is produced from mixtures of rye and wheat flours. The characteristic flavor of rye is liked by many people,

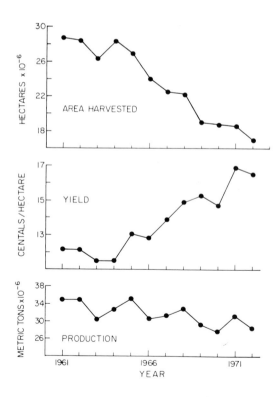

Figure 2. World rye crop area harvested, yield and production.

TABLE I
World cereal crops production (1000 metric tons) for 1963-72[a]

	1963	1964	1965	1966	1967	1968	1969	1970	1971	1972
Wheat	239,581	277,109	267,369	310,107	298,918	331,245	315,190	318,437	353,822	347,603
Rice	255,292	265,588	256,617	254,828	277,488	284,729	293,485	308,767	309,096	295,377
Maize	221,961	215,903	227,814	242,245	266,873	252,701	267,601	261,311	303,612	301,392
Barley	101,826	109,310	106,320	117,272	119,602	131,048	137,025	139,551	151,446	152,238
Oats	47,384	44,314	46,636	48,558	50,825	54,369	55,389	55,557	57,733	51,293
Millet	37,847	40,513	38,674	39,514	42,705	39,682	44,623	47,615	45,337	42,956
Sorghum	36,551	35,356	37,679	41,327	44,575	42,448	45,176	44,648	50,954	46,709
Rye	30,394	32,755	35,453	30,878	31,460	33,355	29,252	27,722	31,693	28,178

[a]Source: FAO (1972).

especially those of eastern Europe. Small quantities of rye flour are used for the production of baked specialty products such as flat bread or rye crisp.

Although used extensively as livestock feed, rye is relatively low on the scale of quality of feed grains. It tends to form a sticky mass in the animals' mouths and can be extremely unpalatable if ergot is present. It is generally used in a mixture with other grains. On occasion, the price of rye is such that it becomes an attractive feed grain in spite of its low feed value.

Substantial quantities of rye grain are used for the production of alcoholic beverages.

Rye is used extensively as a pasture crop. It can be pastured both in the autumn and spring or only in the autumn and cropped in the spring. Occasionally it is grazed in the autumn and used as spring cover crop or ploughed under as green manure for a crop of higher economic value.

Rye straw is fibrous and tough and therefore not used extensively in livestock feed. It is highly desirable, however, for livestock bedding. Small quantities of straw are used in the manufacture of strawboard and paper.

In spite of its many deficiencies, rye will continue to be an important crop because it has a number of advantages over other crops. It is considerably more winter hardy than wheat and will give economical yields on poor sandy soils where no other useful crop can grow. It is grown in many areas that have no

TABLE II

Rye and wheat production by countries for 1972[a]

	1000 Metric Tons	
	Rye	Wheat
U.S.S.R.	9,600	85,800
Poland	8,149	5,147
Germany FED	2,914	6,608
Germany DR	1,900	2,744
U.S.A.	750	42,043
Turkey	740	12,085
Argentina	690	8,100
Czechoslovakia	450	4,220
Austria	402	863
Sweden	363	1,150
Canada	344	14,514
France	331	18,123
All Others	1,545	146,206
World	28,178	347,603

[a]Source: FAO (1972).

alternative crop. It is a good rotational crop because of its ability to combat weeds. In some countries, it is used as a pioneer crop to improve wasteland and sterile soils. In Argentina, it is an important pasture crop; and in southern Australia, it is planted to prevent wind erosion. Its many uses and advantages far outweigh its deficiencies.

VI. INTERNATIONAL TRADE

During the last decade, world trade in rye showed a sharp drop from 1963 to 1964, followed by relatively small fluctuations after 1964 to 1972 (Figure 3, Tables III and IV). In 1972, 2.3% of the total world production was exported. This is a substantial decrease from 5%, the equivalent figure for 1963. Domestic consumption of rye is greatest in countries where the crop has been traditionally grown. Only one major nonproducing country, Japan, has become a major consumer of rye.

The five major exporters (Table III) in the last decade were U.S.S.R., Canada, U.S.A., the German Federal Republic and Sweden. The German Federal Republic was a major exporter of rye in 1971 and 1972, second only to Canada. The U.S.S.R., a major exporter of rye for many years, did not export any in 1972 because of domestic shortages of food and feed grains.

Over the last decade, Japan has gradually become the largest importer of rye (Table IV). Most of this grain in Japan has been used by the animal feed industry; this arose mainly from a price advantage that rye has had in the 1960's over feed grains. Substantial quantities go into the baking industry; rye bread is gaining in popularity in Japan. The other importers in decreasing order of importance are Czechoslovakia, the German Federal Republic, Poland, the German Democratic Republic, and The Netherlands. All but the last-named country are also major producers of rye and are highly dependent on the crop for bread purposes. This need cannot be easily filled by wheat or other cereals because of traditional food preferences and economic reasons. During years of

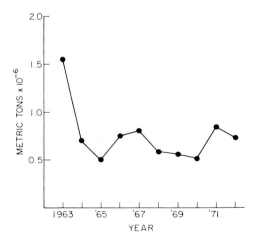

Figure 3. World rye exports.

TABLE III
Major exporters of rye (1000 metric tons) for 1963–72[a]

	1963	1964	1965	1966	1967	1968	1969	1970	1971	1972
Africa	...	...	...	...	...	0.02	0.01	...	...	...
Tanzania	...	...	...	...	...	0.02	0.01	...	...	...
North America	483	293	175	351	308	183	91.4	149	406	416
Canada	113	149	116	243	205	134	66.7	146	270	240
U.S.A.	370	144	59.0	108	103	49.0	24.7	3.0	136	5.0
South America	2.5	112	96.4	4.6	3.4	19.2	13.4	26.4	3.8	2.0
Argentina	2.5	112	96.4	4.6	3.4	19.2	13.4	26.4	3.8	2.0
Asia	0.5	47.7	73.9	22.9	...	...	...	...	0.1	0.1
Turkey	...	47.7	73.9	22.8	...	...	...	...	...	...
Europe	233	66.6	78.1	109	151	146	21.9	272	410	416
Belgium	...	3.7	0.7	3.2	2.1	2.9	3.8	0.8	0.9	10.1
France	9.6	23.3	24.9	18.9	15.9	30.3	30.9	36.8	26.6	62.2
Germany FED	6.5	4.7	3.9	17.6	9.0	3.3	1.2	98.8	218	222
Netherlands	34.0	10.0	8.6	2.0	17.4	26.7	73.5	28.6	44.4	26.2
Poland	...	...	...	19.5	65.5	23.7	68.4	38.7	17.0	30.0
Romania	8.0	17.7	22.7	25.7	31.5	9.0	...	...	...	...
Sweden	18.4	1.2	13.4	16.9	2.7	22.4	27.4	32.6	103	64.2
U.S.S.R.	815	150	36.9	275	336	222	222	172	208	...
World	1,535	670	460	763	799	569	546	620	1,026	662

[a]Source: FAO (1972).

TABLE IV
Major importers of rye (1000 metric tons) for 1963-72[a]

	1963	1964	1965	1966	1967	1968	1969	1970	1971	1972
Africa										
South Africa	1.1	2.5	0.2	...	1.8	1.5	1.1	3.3	1.8	1.4
	1.1	2.5	0.2	...	1.8	1.4	1.1	3.2	1.8	1.4
North America										
U.S.A.	16.8	91.3	29.2	32.5	16.5	25.1	13.5	29.9	1.7	4.0
	16.7	91.3	29.2	32.3	16.5	25.0	13.5	29.8	1.6	3.9
South America										
Venezuela	0.08	0.09	0.2	0.1	0.07	0.04	0.04	0.06	0.5	0.5
	0.08	0.09	0.2	0.08	0.06	0.04	0.04	0.03	0.04	0.05
Asia	9.2	4.4	51.4	73.7	101	66.2	32.1	72.7	160	168
Japan	8.2	4.4	45.9	73.7	101	66.2	28.2	72.7	160	168
Europe	1,522	602	420	641	682	487	503	399	664	514
Belgium	53.6	36.5	36.3	23.5	14.6	12.2	9.2	9.4	8.2	6.2
Czechoslovakia	61.5	40.9	32.1	53.4	176	120	85.0	77.4	215	120
Finland	48.9	43.9	38.2	6.0	67.4	23.7	4.1	10.2	19.8	21.4
Germany DR	282	101	1.0	111	50.5	35.4	91.6	49.2	40.0	30.0
Germany FED	182	53.0	21.3	42.2	83.6	60.7	118	71.7	42.7	61.8
Netherlands	268	165	115	110	46.6	37.0	41.0	23.5	25.8	33.0
Poland	409	...	...	70.3	...	63.9	71.7	55.8	111	60.0
Sweden	118	82.8	49.2	72.5	40.2	17.9	3.9	5.1	0.02	0.01
World	1,549	700	501	747	802	580	550	505	828	728

[a]Source: FAO (1972).

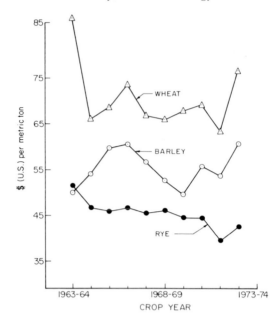

Figure 4. Average annual price in Minneapolis, Minn., of rye, barley, and wheat (Jiler, 1973).

exceptionally good yields, the production in these countries meets the domestic requirements. Accordingly, their imports fluctuate markedly from year to year depending on domestic production. For example, imports by the German Federal Republic during the period 1963-72 ranged from a low of 21,300 metric tons in 1965 to 182,000 metric tons in 1963. Poland, which normally imports between 50,000 and 100,000 metric tons of rye, did not import any in 1964, 1965, and 1967. International trade in rye is, therefore, highly dependent on the production of the major consuming countries.

The price of rye has gradually decreased over the last ten years (Figure 4). In any one year, the price fluctuates widely depending on supply and demand. Generally, rye is priced substantially below wheat on the international market; on occasion, however, scarcity has brought rye prices close to those of wheat. During times when production has exceeded demand and the price of rye has been substantially below the price of barley, considerable quantities of rye have been used in animal feed.

Rye from the various exporting countries differs considerably in price and in quality. From any one source, price differences for varieties or grades of rye are relatively small. The price spread between rye grain, flour, and bread is considerably smaller than the equivalent spread for wheat. Rye bread prices in the major consuming countries are always lower than wheat bread prices, and rye bread is generally a staple food for the poorer class of people. In many countries where bread is baked from a mixture of rye and wheat flours, a low-grade wheat flour is used. This further lowers the price of rye bread relative to the standard wheat product. Although rye is inferior in many ways to the major cereal crops, wheat, rice, and maize, it will continue to be an important crop to the farmer because of its extreme hardiness and ability to grow in low-fertility soil and to the consumer because of its relatively low price.

LITERATURE CITED

CARLETON, M. A. 1919. The small grains. Macmillan: New York.

DEODIKAR, G. B. 1963. Rye, *Secale cereale* Linn. Indian Council Agr. Res.: New Delhi.

FAO (Food and Agriculture Organization). 1972. Production yearbook: 26.

JILER, H. 1973. Commodity year book. Commodity Res. Bureau, Inc.: New York.

KUCKUCK, J. 1937. Zur Entstehung und Abstammung des Roggens. [P.B.A. 8: 487.] Z. Gesamte Getreidew. 24: 131-132.

LEONARD, W. H., and MARTIN, J. H. 1963. Rye. In: Cereal Crops, ed. by W. H. Leonard and J. H. Martin. Macmillan: New York.

POPOV, A. 1939. Investigations on the botanical diversity and gapping in our native rye. Annu. Univ. Sofia 17: 523 (quoted in Deodikar, 1963).

SCHEIBE, A. 1935. Die Verbreitung von Unkrautroggen und Taumelloloch in Anatolien. (Mit Bemerkungen zum Roggen-Abstrammungsproblem.) Angew. Bot. 17: 1-22.

CHAPTER 2

CYTOGENETICS, PLANT BREEDING, AND AGRONOMY

L. E. EVANS
G. J. SCOLES
Department of Plant Science
The University of Manitoba, Winnipeg, Manitoba, Canada

I. CYTOGENETICS

The genus *Secale* L. (cultivated and wild ryes), together with the genera *Triticum* L., *Aegilops* L., *Agropyron* Gaertn., and *Haynaldia* Schur., is classified in the sub-tribe *Triticinae* of the tribe *Triticeae* within the grass family, *Gramineae*. Members of the *Triticeae* have a basic chromosome number of n = 7. Polyploidy, the multiplication of the normal diploid number of chromosomes, has played a major role in the evolution of most of these genera; but in the case of *Secale* and *Haynaldia,* naturally occurring polyploids are unknown (Avdulov, 1936) and all species are characterized by a diploid chromosome number of 2n = 14.

A review (Gotoh, 1931) of early cytological work in *Secale* reveals much confusion in chromosome numbers. Němec (1910) and Nakao (1911) reported haploid numbers of 12 and 8, respectively. Sakamura (1918) appears to have given the first correct report of n = 7, but confusion continued (Gotoh, 1924; Kihara, 1924) until Darlington (1933) and Hasegawa (1934) reported a chromosome number of 2n = 14 plus variable numbers of B chromosomes.

Despite the low chromosome number and the relatively large size of the chromosomes, the exact karyotype of *S. cereale* was subject to much disagreement. Oinuma (1953) explained the earlier confusion after a study of several European and Oriental cultivars. Although he was able to distinguish the seven pairs of chromosomes in each cultivar, he noted significant karyotype differences. Similar variation has been reported between different inbred lines by Bose (1956, 1957) and Heneen (1962). Bhattacharya and Jenkins (1960) presented a karyotype of the cultivar Dakold in which the seven chromosomes were distinguished on the basis of length, arm ratio, and the occurrence and location of secondary constrictions.

Subsequent to Sears' (1966) arrangement of the 21 chromosomes of common wheat into 3 genomes with 7 homoeologous groups of 3 chromosomes each, it became apparent that the rye genome was homoeologous to the 3 wheat genomes and that the rye chromosomes could be assigned to the specific homoeologous groups and numbered 1R through 7R (Riley, 1965; Jenkins, 1966; Sears, 1968; Lee *et al.*, 1969; Gupta, 1969, 1971). Darvey (1973) outlined a tentative placement of the rye chromosomes into their homoeologous groups based on the available literature (Table I). It is apparent from these studies that translocations have occurred in rye, complicating the establishment of homoeologies, and that *S. montanum* may represent the primitive rye karyotype.

Studies of interspecific hybrids (Schiemann and Nurnberg-Krüger, 1952; Khush and Stebbins, 1961; Stutz, 1972) indicated that the species *S. montanum, S. africanum, S. silvestre,* and *S. vavilovii* differ from *S. cereale* by two, and in the case of *S. silvestre* by perhaps three, reciprocal translocations. Khush (1962) also found translocation differences between the related species. Stutz (1972), on the basis of cytological analyses of interspecific hybrids, morphology, and chemotaxonomy, recognized six *Secale* species: *S. montanum, S. anatolicum,* and *S. africanum* belonging to the *Kuprijanovii* (Roshev.) series, all having the same chromosome configuration; *S. cereale* and *S. vavilovii* belonging to the *Cerealia* (Roshev.) series which differ from the others by three translocations; and *S. silvestre,* an annual species like the *Cerealia,* which has the *Kuprijanovia* configuration and constitutes the third series *Silvestria* (Roshev.).

The chromomeric structure of rye chromosomes, first reported by Shmargon (1938) and later studied extensively by Lima de Faria (1952), enabled the latter to map each of the chromosomes of rye and to identify them on this basis. Heterochromatin staining procedures (Sarma and Natarajan, 1973; Verma and Rees, 1974; Gill and Kimber, 1974; Darvey and Gustafson, 1975) have been used to identify the seven rye chromosomes and characterize them as to heterochromatin pattern (Figure 1).

Meiosis of rye was studied in detail by Gurney (1931), Darlington (1933), and Sax (1935) and was generally found to be regular. Darlington (1933) found an average of 2.4 chiasmata per bivalent with the number being partially dependent

TABLE I

Tentative placement of rye chromosomes into their homoeologous groups[a]

Homoeologous Group	Chinese Spring × Imperial (Sears' Designations)	Holdfast × King II (Riley's Designations)	Kharkov & Dakold (Evans' & Jenkins' Designations)
1R	E	V	VII
2R	B	III	II
3R	G	VI	I
4R/7R	D	IV	V
5R	A	I	VI
6R	F	II	IV
7R/4R	C	VII	III

[a]Source: Darvey (1973).

on chromosome length. Within the rye spike, the first pollen mother cells to undergo meiosis occur in florets about one-third of the way down from the tip of the spike, and meiosis proceeds progressively up and down from this point. This gradient is reflected in the pattern of anthesis which progresses in a similar manner.

Lamm (1936) noted that inbreeding of rye, a naturally outbreeding species, had an adverse effect on the regularity of meiosis. These results were later substantiated by Müntzing and Akdik (1948) who found, in addition, a positive correlation between chromosome pairing and fertility in inbred lines. Rees and

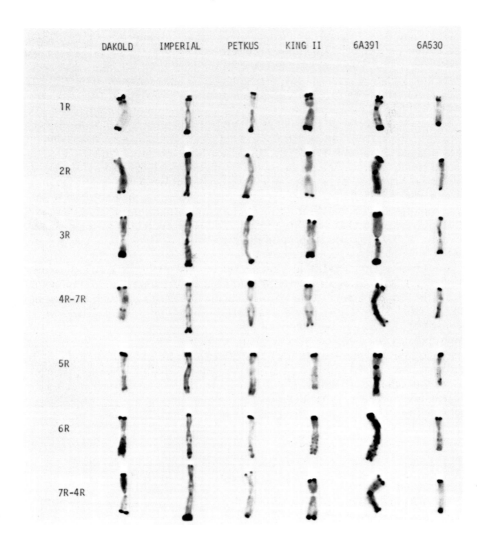

Figure 1. Leishman C-banding of the rye chromosomes of four additional series (Dakold, Imperial, Petkus, and King II) to wheat, and two triticales (6A391 and 6A530). After Darvey and Gustafson, 1975.

co-workers (Rees, 1955; Rees and Thompson, 1956; Jones and Rees, 1964; Hazarika and Rees, 1967) used inbred lines of different meiotic stability to investigate the control of meiotic characters. They concluded that the differences in mean chiasmata frequency were a result of genotypic control of the variation of chiasmata frequency among pollen mother cells within a plant. Other factors, such as the distribution of chiasmata among bivalents within any pollen mother cell and the method of disjunction of quadrivalents (adjacent or alternate), were also under genotypic control. In an outbreeding population, a genetic balance is preserved between these genes; but on inbreeding, this is disturbed as the genes become homozygous. As a result, inbred lines show irregularities during their meiotic division.

Although B, or accessory, chromosomes occur in many plant and animal species, they are especially common in genera belonging to the *Gramineae* (Jones and Rees, 1967). They are very frequent in the wild or primitive forms of rye, particularly those from Asia (Müntzing, 1957). B chromosomes are unlike the normal A chromosomes in that their absence is not deleterious and their number varies within a population.

Müntzing and Lima de Faria (1949) described a standard B chromosome type plus five other types (telocentrics, isochromosomes, etc.) derived from it by means of chromosome breakage. The B chromosomes do not pair with A chromosomes at meiotic metaphase, but homologous B chromosomes may pair with each other although usually with reduced frequency as compared to A chromosomes (Müntzing, 1945). A peculiar property of B chromosomes is that they undergo directional nondisjunction at the first maturation division of the pollen grains. This may result in increased numbers of B chromosomes in the progeny of B chromosome-carrying plants (Müntzing, 1966; Jain, 1960).

There is considerable disagreement as to the value of B chromosomes. Rhoades and Dempsey (1972) consider their presence to be a parasitic one, whereas Jones and Rees (1967) argue that under certain environments, B chromosomes offer some advantages to the plant that carries them. Darlington (1937) and Mori (1948) have found differences in the frequency of B chromosomes carried by plants of the same genotype grown in different environments for a period of time. Müntzing (1943) found that reproductive characters of the plant were adversely affected by increasing numbers of B chromosomes. Jones and Rees (1967) investigated the effect of B chromosomes on meiotic characters and contend that they affect the pattern of chiasmata distribution and hence affect the rate of recombination and release of variability. John and Jones (1970) found an interaction between B chromosomes and histone proteins, implicating B chromosomes in the control of genetic activity through changes in histone level. (Histones are thought to act as blanket repressors of genetic activity.)

II. GENETICS

The facts that rye is a cross-pollinated species and that inbred lines usually lack vigor have restricted genetic analyses. However, the recent production of rye chromosome additions to or substitutions into wheat has permitted specific characteristics to be attributed to specific rye chromosomes. The genetic analyses of some of the more important characteristics of rye are discussed in the following sections.

A. Growth Habit

The inheritance of perennial versus annual growth habit has been studied using interspecific hybrids due to the lack of perennial forms of *S. cereale*. In such crosses, perennial habit is at least partially dominant (Tschermak, 1906; Deržhavin, 1935; Riley, 1955). The progeny from such crosses express a complete array of growth habits preventing easy determination of the number of factors involved (Ossent, 1930; Stutz, 1957).

B. Winter Versus Spring Habit

Studies by Tschermak (1906) and Purvis (1939) showed spring habit to be dominant and simply inherited. Wexelsen (1969), in a study of both diploid and tetraploid cultivars, found spring habit to be dominant and controlled by three dominant genes in the diploids; two dominant genes were indicated in the tetraploids.

Several reports of U.S.S.R. investigators (Šerstova, 1959; Ostanin, 1959; Medvedev, 1965; Stroun *et al.*, 1960) detailed the success in transforming spring types to winter and *vice versa* by reversing planting dates.

C. Ear Characteristics

Rachis fragility is a characteristic of wild rye which is detrimental to a domesticated crop and has been eliminated from *S. cereale*. It is a dominant character controlled by a single gene (Kostoff, 1937; Jermoljev, 1941; Kranz, 1963). Earlier work indicated that brittle rachis and perennial habit were closely linked (Ossent, 1930; Duka, 1938).

An additional factor, a monogenic recessive, causing both rachis and stem brittleness, has been reported by several workers (Brewbaker, 1926; Dumon, 1932; Lada, 1933; Sybenga and Prakken, 1962). This monogenic recessive apparently occurred spontaneously in *S. cereale* and is distinct from the previously mentioned dominant gene.

Ear branching has been studied by many (Thompson, 1922; Molotkovskii, 1950; Borojević, 1953; Szentivány, 1954; Lekčsinkaja and Vioncek, 1955; Necas, 1961) and the consensus is that it is controlled by a single recessive gene. It is very sensitive to environmental conditions with maximum expression requiring ideal conditions.

Stutz (1958) reported a mutant having three spikelets per rachis node inherited as a simple recessive except in crosses to *S. montanum* when it appeared as a dominant.

D. Anthocyanin Pigmentation

Anthocyanin pigments may occur in the aleurone layer of the grain, the coleoptile, first leaf, stem base, and nodes, the upper internode, and in the anthers (Sybenga and Prakken, 1962). Rümker (1912) and Steglich and Pieper (1922) found the anthocyanin-containing green seed color dominant to the anthocyanin-free yellow seed type. Although the inheritance was monogenic, it was modified by the independently inherited epidermis color.

Dumon (1932, 1938) observed a xenia effect on seed color and a pleiotropic effect of seed color on seed size. He proposed a complementary two-gene system controlling aleurone color. Neumann and Pelshenke (1954) constructed a seed color chart based on modifications caused by the color of the pericarp and testa.

Watkins and White (1964) concluded that grain color was controlled by two complementary genes A and B and that anthocyanin pigmentation in other organs was controlled by the action of gene A and an additional gene R. They found genes A and B to be linked with a recombination value of 5.6%.

E. Incompatability and Self-Fertility

Lundquist, in a series of reports (1954, 1956, 1957, 1958a,b) presented a comprehensive coverage of the incompatability mechanism in rye. He confirmed earlier results (Peterson, 1934; Heribert-Nilsson, 1953) that two independent multi-allelic loci S and Z controlled incompatability.

Kuckuck and Peters (1967) intercrossed self-sterile and self-fertile cultivars and selected and intercrossed self-fertile progeny. They were able to improve the self-fertility and found it dominant to self-sterility.

Wricke (1969) questioned the relationship of the incompatability system and fertility. Although it is still not entirely clear, there is some evidence that two separate systems are involved.

F. Chlorophyll Deficiencies

Numerous chlorophyll deficiencies following inbreeding have been reported. Brewbaker (1926) classified these deficient types into four categories: white, virescent white, yellow-green, and striped. They are all caused by recessive factors.

Dumon (1953, 1962) and Dumon and Laeremans (1963) proposed four loci, each with multiple alleles, as controlling these deficient types. The normal green genotype would be $C_1C_2W_1W_2$ whereas various recessive combinations could explain the chlorophyll-deficient types.

G. Dwarfing

As in other cereal crops, dwarfing genes have been sought as a tool in producing improved varieties. Inbreeding of rye usually results in reduced height as a result of inbreeding depression. Sybenga and Prakken (1962) analyzed two dwarf inbred lines and found them to be controlled by different single recessive genes designated d_1 and d_2.

Fedorov *et al.* (1970) and Kobyljanskii (1972, 1973) have reported three different types of dwarfs, two of which are simply inherited recessives controlled by the genes ct and hl, respectively, and a more complex recessive inheritance in the cultivar Petkuser Kurzstroh.

H. Protein Content and Amino Acid Composition

The addition to or substitution of rye chromosomes into wheat has permitted analyses of the effects of different rye chromosomes on an otherwise uniform wheat genotype. Shepherd (1968) and Riley and Ewart (1970) utilized the seven King II rye additions to Holdfast wheat to study the influence of the rye chromosomes on protein content. Shepherd (1968) found chromosome 1R to be involved in the genetic control of endosperm protein. Riley and Ewart (1970) found that chromosomes 3R, 4R, 6R, and 7R increased total protein by 3 to 4.5%. Chromosome 5R increased lysine by 8.7% and cystine by 10.7%. Chromosome 6R increased proline by 9.1% and decreased aspartic acid content by 8.6%. The other chromosomes had some influence on other amino acids.

Jagannath and Bhatia (1972) found that substitutions of chromosome 2R of Imperial rye for homoeologous group 2 chromosomes of Chinese Spring increased protein content by as much as 7%.

I. Other Characters

Substitution and addition of rye chromosomes to wheat have also been used to locate genes for various characters on particular rye chromosomes. O'Mara (1946) located the gene for hairy-neck of rye on the chromosome now designated as 5R. Chapman and Riley (1955), Riley and Chapman (1958), Evans (1959), and Evans and Jenkins (1960) studied various rye additions to wheat and outlined the morphological effects of each addition.

Darvey (1973) has studied the influence of each rye chromosome on kernel shrivelling and has concluded that all rye chromosomes, except possibly 2R, are involved.

Riley and Macer (1966) investigated the chromosomal location of factors for resistance to wheat pathogens in rye and found that resistance to *Puccinia striiformis, Erysiphe graminis,* and *Cercosporella herpotrichoides* was associated with specific rye chromosomes. This finding may be useful in wheat breeding.

J. Cytoplasm Effects

Frimmel (1939), Wellensiek (1948), and Jain (1960) reported a maternal effect on grain characteristics. Dumon (1957) reported a chlorophyll deficiency which was cytoplasmically inherited and also suggested that straw length may be cytoplasmically inherited.

Alloplasmic rye (rye nucleus in wheat cytoplasm) and wheat and triticale in rye cytoplasm have been produced and studied (Lein, 1948; Lacadena and Pérez, 1973). Maan and Lucken (1971) found alloplasmic wheat (wheat in rye cytoplasm) to be male sterile, partially female sterile, and low in vigor. However, the addition of one or a pair of rye chromosomes restored fertility and vigor. Maan (1973) reported on the fertility-restoring properties of the various rye chromosomes and found 3R to be most effective in restoring fertility whereas 6R and 7R were totally ineffective. Sánchez-Monge and Soler (1973) evaluated triticales having rye and wheat cytoplasm and concluded that those possessing rye cytoplasm were inferior. Rimpau (1973) and Rimpau et al. (1973) investigated the influence of rye cytoplasm on rye chromosome additions or substitutions in alloplasmic wheat and found a change in plant morphology, meiotic regularity, and rye chromosome transmission as compared to similar material in wheat cytoplasm.

Male sterile forms of rye were investigated by Schnell (1960), Čekhovskaja (1965), and Kobyljanskii (1969). The latter attributed the sterility to cytoplasmic factors. Geiger and Schnell (1970) isolated cytoplasmic male sterile lines from a cross between Argentine Pampa rye and a German inbred line. They have since released these lines along with the required restorer and maintainer line. These stocks permit the evaluation of hybrid rye production.

III. PLANT BREEDING

Because of the relatively small acreage and its outbreeding nature, little effort has gone into rye breeding. Improvements have been made in seed size, yield, winter hardiness, and plant height by selection within heterogeneous populations.

Disease is not a major problem in rye, with the possible exception of ergot caused by the organism *Claviceps purpurea.* No source of resistance has yet been identified within the genus, so breeding for ergot resistance is not possible.

Very little emphasis has been placed on rye quality, either for human consumption or as a livestock feed. As a feedstuff, palatability problems and growth-inhibiting factors have been associated with rye and have restricted its use. As yet, the compounds causing the feeding problems have not been positively identified; the pentosans, trypsin inhibitors, and alkylresorcinols have been suggested as possibilities.

Production of distilled alcoholic beverages is the next most important use of rye after food and feed uses. For this purpose, a light-colored grain is preferred. The maintenance of a specific seed color requires that cultivars of differing colors be isolated during production. Although the methods for the improvement of an outbreeding species have been fully documented and used in corn, the same procedures have rarely been applied to rye.

Deodikar (1963) has classified the breeding procedures for rye as follows: intraspecific breeding, interspecific, intergeneric, and polyploidy and mutation. Use of each procedure will be discussed briefly.

INTRASPECIFIC BREEDING

Intraspecific breeding refers to hybridization of lines or cultivars within the species. This is the most commonly used breeding procedure for rye.

Sprague (1938) recommended continuous selection in rye breeding. The cultivar Raritan was derived from a population obtained by allowing 10 strains of rye to intercross freely for 2 years. The progeny were then grown as single plants. Of these, 216 were selected and yield-tested over a number of years. The 98 best lines were bulked to establish the cultivar. Similar techniques have been used in the production of the current Canadian fall rye cultivars.

Recurrent and reciprocal recurrent types of selection have both been suggested for rye. Ferwerda (1962) obtained a synthetic variety after one cycle of recurrent selection that outyielded the standard cultivar Petkus by 20%.

Bauer (1972) utilized reciprocal recurrent selection to obtain parents with good combining ability. He suggested the use of clonal propagation to produce sufficient seed after three or five cycles of selection.

Ferwerda (1962) and Walther (1959) both used self-pollination to maintain the strains used in polycross tests and for the production of synthetics.

A germ plasm source population has been established by Qualset and Hoskinson (1966) utilizing 43 different North American sources of the Balbo cultivar and 10 other North American cultivars. The 10 cultivars were used as a pollen source only, and the open-pollinated seed of the Balbo strains was composited. This source population hopefully will permit selection for such characteristics as winter hardiness, disease resistance, and agronomic suitability.

Rümker and Leidner (1914) documented the existence of hybrid vigor in rye almost as early as in corn. However, it is only recently that the stocks necessary for hybrid rye production have become available. Geiger and Schnell (1970) and Kobyljanskii (1969) have reported two sources of cytoplasmic male sterility and the sources of fertility restoration.

INTERSPECIFIC BREEDING

Several investigators (Tschermak, 1913; Ossent, 1930; Deržhavin, 1935; Duka, 1938; Dierks and Reimann-Phillipp, 1966) have crossed *S. cereale* with other rye species, particularly *S. montanum,* in an attempt to improve winter hardiness, drought resistance, and disease resistance. Hondelmann and Snejd (1973) transferred the perennial habit to cultivated rye and obtained fertility and yield equal to annual cultivars.

Interspecific breeding has also been used to produce rye with high protein content. Focke (1956) found that *S. montanum* had twice the protein content of *S. cereale* and their hybrid was similar in content to *S. montanum* but with a better amino acid balance than either parent. Kobyljanskii (1968) and Kotvics (1970) also found the wild rye species to be high in protein. Kotvics (1970) was able to obtain high protein lines in the progeny of hybrids with cultivated rye, having increased biological protein value.

INTERGENERIC BREEDING

Secale has been crossed with the other five genera in the Triticinae sub-tribe but little or no F_1 pairing occurred and the hybrids were sterile. Successful crosses have been obtained between *Secale cereale* and *Hordeum jubatum* (a wild barley) and *Elymus arenarius* but the resulting F_1 plants were sterile. Although some fertile amphiploids have resulted from natural or artificial chromosome doubling, the only intergeneric hybrid of practical significance is triticale resulting from *Secale-Triticum* hybrids.

Intergeneric hybrids have been of no significance to rye breeding but may be of value in producing cytoplasmic male sterile lines (Lacadena 1967) for use in hybrid rye breeding.

POLYPLOIDY AND MUTATION

Müntzing (1951) has reviewed the rather intensive effort made in producing auto-tetraploid rye. In general, auto-tetraploid rye is superior to diploid rye in kernel size, protein content, and baking quality; but its yield is inferior due to sterility.

Subsequent to Dorsey's (1936) production of tetraploid Rosen rye, several other tetraploid cultivars have been produced. Some have been released in Europe for commercial production but they have not been of economic significance.

IV. AGRONOMY

Rye is the most widely adapted of the cereals, capable of being grown within the arctic circle in Scandinavia and in the southerly latitudes of Chile. Rye is also cultivated at altitudes as high as 4,300 m in the Himalayas. This wide distribution is due to its extreme winter hardiness and ability to grow on very marginal soils. Rye is the most drought-resistant of the cereals due to its extensive root system and its inherent ability to adjust its maturity to moisture conditions (Deodikar, 1963).

Rye is also able to tolerate a range of soil acidity and is widely grown in areas that are too acid for wheat production. Although rye is usually produced on poorer soils, it will respond to good fertility and moisture conditions; yields of 1.5 metric tons per acre have been recorded in the United Kingdom.

In North America, most of the rye area is seeded to winter cultivars due to their winter hardiness and ability to utilize early spring moisture. Winter rye cultivars vary considerably in hardiness but the most hardy rye cultivars grown in the northern United States and Canada are considerably hardier than the most hardy wheat cultivars. In much of this area, fall rye is the only crop with sufficient hardiness for safe production as a winter crop.

Fall-sown rye is frequently grazed in the late fall and early spring and then left to produce a grain crop. This practice requires considerable care as to the timing of grazing or serious yield reduction may occur.

The extreme winter hardiness of rye and its ability to grow in low-fertility soils make it an attractive crop for some areas. These attributes could be the important features which will prompt efforts to overcome other shortcomings by breeding varieties with improved quality and agronomic characteristics.

LITERATURE CITED

AVDULOV, N. P. 1936. Karyosystematic investigations in the family *Gramineae.* Bull. Appl. Bot. Genet. Plant Breed. Suppl. 43: 428.

BAUER, F. 1972. Results of two cycles of reciprocal recurrent selection in rye. [P.B.A. 42: 7485.] Kiserletugyi Kozlem. 62A (1/3): 41-52.

BHATTACHARYA, N. K., and JENKINS, B. C. 1960. Karyotype analysis and chromosome designations for *Secale cereale* L. "Dakold". Can. J. Genet. Cytol. 2: 268-277.

BOROJEVIĆ, S. 1953. Branched ears in rye. [P.B.A. 27: 2875.] Poljopr. Znanst. Smot. 15: 75-87.

BOSE, S. 1956. Aberrants in nuclear chromosomes of inbred rye. Hereditas 42: 263-292.

BOSE, S. 1957. Aberrations in the nucleolar chromosomes of inbred rye. II. Size variation in inbred lines and population plants. Hereditas 43: 621-643.

BREWBAKER, H. E. 1926. Studies of self-fertilization in rye. Univ. Minn. Agr. Exp. Sta., Tech. Bull. 40.

ČEKHOVSKAJA, E. S. 1965. Sterile forms of winter rye. [P.B.A. 36: 3777. Selek. Semenovod. 30(5): 53-54.

CHAPMAN, V., and RILEY, R. 1955. The disomic addition of rye chromosome II to wheat. Nature 175: 1091.

DARLINGTON, C. D. 1933. The origin and behaviour of chiasmata VIII. *Secale cereale.* Cytologia 4: 444-452.

DARLINGTON, C. D. 1937. Cited by Deodikar (1963).

DARVEY, N. L. 1973. Genetics of seed shrivelling in wheat and triticale. Proc. 4th Int. Wheat Genet. Symp., Columbia, Mo. p. 155-159.

DARVEY, N. L., and GUSTAFSON, J. P. 1975. Identification of rye chromosomes in wheat-rye addition lines and triticale by heterochromatin bands. Crop Sci. 15:239-263.

DEODIKAR, G. B. 1963. Rye. Indian Counc. Agr. Res.: New Delhi. p. 152.

DERŽHAVIN, A. I. 1935. Further data on the perennial rye *Secale kuprijanovii* Grossh, and its agricultural value. [*P*.B.A. 6: 868.] Bull. Appl. Bot. Leningrad: Ser. A(14): 159-165.

DIERKS, W., and RIEMANN-PHILLIPP, R. 1966. Breeding a perennial form of rye as a possibility of improving the technique of rye breeding and of creating a multiple purpose rye for use as green fodder and grain. Z. Pflanzenzüchtg. 56: 343-368.

DORSEY, E. 1936. Induced polyploidy in wheat and rye. J. Hered. 27: 155-160.

DUKA, S. Kh. 1938. Methods of breeding cultivated perennial rye. [P.B.A. 9: 719.] Selek. Semenovod. 8-9: 18-20.

DUMON, A. G. 1932. Genetical researches on rye (*Secale cereale* L.). [P.B.A. 3: 404.] Meded. Sta. Plantenvered. Héverlee-Leuven: No. 5. p. 19.

DUMON, A. G. 1938. A case of dominant and recessive brown in *S. cereale* L. [P.B.A. 9: 718.] Agricultura (Louvain) 41: 190-196.

DUMON, A. G. 1953. The heredity of chlorophyll deficiencies in *S. cereale* L. [P.B.A. 27: 1505.] Proc. 9th Int. Congr. Genet., Bellagio (Como), August 26-31. p. 1221-1225.

DUMON, A. G. 1957. Scientific research in agriculture 1947-1956: Part 1. P.B.A. 29: 3540.

DUMON, A. G. 1962. Analysis of the heritable properties of rye (*S. cereale* L.). [P.B.A. 32: 3217.] Studkring. Plant Veredel. 70: 966-982.

DUMON, A. G., and LAEREMANS, R. 1963. The inheritance of chlorophyll in *S. cereale* L. [P.B.A. 33: 4404.] Agricultura (Louvain) 11: 91-105.

EVANS, L. E. 1959. *Agropyron* and *Secale* chromosome additions to common wheat. Ph.D. Thesis. University of Manitoba. p. 46.

EVANS, L. E., and JENKINS, B. C. 1960. Individual *Secale cereale* chromosome additions to *Triticum aestivum*. I. The addition of individual "Dakold" fall rye chromosomes to "Kharkov" winter wheat and their subsequent identification. Can. J. Genet. Cytol. 2: 205-215.

FEDOROV, V. S., SMIRNOV, V. G., and SOSNIHINA, S. P. 1970. The genetics of rye. X. The inheritance of dwarfness. [P.B.A. 40: 7295.] Genetika 6(3): 5-17.

FERWERDA, F. P. 1962. Results of one cycle of recurrent selection in rye. Euphytica 11: 221-228.

FOCKE, R. 1956. On the crude-protein content of *S. cereale* and *S. montanum* and the percentage proportion of some amino acids. [P.B.A. 26: 2281.] Zuchter 26: 40-41.

FRIMMEL, F. 1939. Beitrag zur Xenianfrage bei Roggen. Zuchter 11: 301-307.

GEIGER, H. H., and SCHNELL, F. W. 1970. Cytoplasmic male sterility in rye. Crop Sci. 10: 590-593.

GILL, B. S., and KIMBER, G. 1974. The Giemsa C-banded karyotype of rye. Proc. Nat. Acad. Sci. U.S.A. 71: 1247-1249.

GOTOH, K. 1924. Über die chromosome zahl von *Secale cereale* L. Bot. Mag. (Toyko) 38: 135-152.

GOTOH, K. 1931. Further investigations on the chromosome number of *Secale cereale* L. Jap. J. Genet. 7: 172-182.

GUPTA, P. K. 1969. Studies on transmission of rye substitution gametes in common wheat. Indian J. Genet. Plant Breed. 29: 163-172.

GUPTA, P. K. 1971. Homoeologous relationship between wheat and rye chromosomes. Present status. Genetica 42: 199-213.

GURNEY, H. C. 1931. The cytology of rye. Aust. J. Exp. Biol. Med. Sci. 8: 241-254.

HASEGAWA, N. 1934. A cytological study on 8-chromosome rye. Cytologia 6: 68-77.

HAZARIKA, M. H., and REES, H. 1967. Genotypic control of chromosome behaviour in rye. X. Chromosome pairing and fertility in auto-tetraploids. Heredity 22: 317-332.

HENEEN, W. K. 1962. Chromosome morphology in inbred rye. Hereditas 48: 182-200.

HERIBERT-NILSSON, H. 1953. Über die Entstehung der Selbfertilität beim Roggen. Hereditas 39: 236-256.

HONDELMANN, W., and SNEJD, J. 1973. Perennierender Kultur-Roggen. I. Gegenwärtiger stand der Züchtung. Z. Pflanzenzüchtg. 69: 89-101.

JAGANNATH, D. R., and BHATIA, C. R. 1972. Effect of rye chromosome 2 substitution on kernel protein content of wheat. Theor. Appl. Genet. 42: 89-92.

JAIN, S. K. 1960. Cytogenetics of rye (*Secale* spp.). Bibliogr. Genet. 19: 1-86.

JENKINS, B. C. 1966. *Secale* additions and substitutions in common wheat. Proc. 2nd Int. Wheat Genet. Symp., Hereditas Suppl. Vol. 2. p. 301-312.

JERMOLJEV, E. 1941. The brittleness of rye. Z. Pflanzenzüchtg. 24: 59-70.

JOHN, P. C. L., and JONES, R. N. 1970. Molecular heterogeneity of soluble proteins and histones in relationship to the presence of B-chromosomes in rye. Exp. Cell Res. 63: 271-276.

JONES, G. H., and REES, H. 1964.

Genotypic control of chromosome behaviour in rye. VIII. The distribution of chiasmata within pollen mother cells. Heredity 19: 719-730.

JONES, R. N., and REES, H. 1967. Genotypic control of chromosome behaviour in rye. IX. The influence of B-chromosomes on meiosis. Heredity 22: 333-347.

KHUSH, G. S. 1962. Cytogenetic and evolutionary studies in *Secale*. II. Interrelationships in the wild species. Evolution 16: 484-496.

KHUSH, G. S., and STEBBINS, G. L. 1961. Cytogenetic and evolutionary studies in *Secale*. I. Some new data on the ancestry of *S. cereale*. Amer. J. Bot. 48: 723-730.

KIHARA, H. 1924. Cytologischen und genetische studien vei wichtigen getreidearten mit besonderer Rücksicht auf das verhalten der chromosomen und die sterilität in den bastarden. Mem. Coll. Sci. Univ. Kyoto, Ser. B. 1: 1-200.

KOBYLJANSKII, V. D. 1968. Initial material of rye for breeding for grain quality. [P.B.A. 39: 2238.] Selek. Semenovod. 5: 21-23.

KOBYLJANSKII, V. D. 1969. Cytoplasmic male sterility in diploid rye. [P.B.A. 40: 679.] Vestn. Sel'Skokhoz. Nauki 6: 18-22.

KOBYLJANSKII, V. D. 1972. The genetics of the dominant factor for short straw in rye. [P.B.A. 43: 1824.] Genetika 8: 12-17.

KOBYLJANSKII, V. D. 1973. Genetic properties of short straw in rye. [P.B.A. 44: 1570.] Tr. Prikl. Bot., Genet. Selek. 49(3): 97-108.

KOSTOFF, D. 1937. Interspecific hybrids in *Secale* (Rye). I. *Secale cereale × Secale ancestrale, S. cereale × S. vavilovii, S. cereale × S. montanum* and *S. ancestrale × S. vavilovii* hybrids. [P.B.A. 7: 1238.] C. R. (Dokl.) Acad. Sci. U.S.S.R. 14: 213-214.

KOTVICS, G. 1970. Investigations on increasing the protein content of *S. cereale* L. In: Protein growth by plant breeding, ed. by A. Balint. [P.B.A. 41: 6920.] Akad. Kiado: Budapest.

KRANZ, A. R. 1963. Die anatomischen, ökologischen und genetischen Grundlagen der Ährenbruchigkeit des Roggens. Beitr. Biol. Pflanz. 38: 445-472.

KUCKUCK, H., and PETERS, R. 1967. Genetische Untersuchungen über die Selbstfertilität bei *Secale vavilovii* Grossh. und *Secale cereale* L. var. Dakold in Hinblick auf Probleme der Züchtung und Phylogenie. Z. Pflanzenzüchtg. 57: 167-188.

LACADENA, J. R. 1967. Alloplasmic rye. Cytogenetic considerations about its obtention. Port. Acta Biol. 10: 129-142.

LACADENA, J. R., and PÉREZ, M. 1973. Cytogenetical analysis of the interaction between *Triticum durum* cytoplasm and *Secale cereale* nucleus at the diploid and tetraploid nuclear levels. Proc. 4th Int. Wheat Genet. Symp., Columbia, Mo. p. 355-360.

LADA, P. 1933. The genetics of brittle rye. [P.B.A. 5: 351.] Bull. Acad. Pol. Sci. Lettres 1933: 183-193.

LAMM, R. 1936. Cytological studies on inbred rye. Hereditas 22: 217-240.

LEE, Y. H., LARTER, E. N., and EVANS, L. E. 1969. Homoeologous relationship of rye chromosome IV with two homoeologous groups from wheat. Can. J. Genet. Cytol. 11: 803-809.

LEIN, A. 1948. Über alloplasmatische Roggen (Roggen mit Weizenplasma). Zuchter 19: 101-108.

LEKCSINKAJA, J., and VIONCEK, J. I. 1955. Conditions promoting branched ears in rye. [P.B.A. 25: 2978.] Agrobiologiya 2: 78-82.

LIMA DE FARIA, A. 1952. Chromosome analysis of the chromosome complement of rye. Chromosoma 5: 1-68.

LUNDQUIST, A. 1954. Studies on self sterility in rye. Hereditas 40: 278-294.

LUNDQUIST, A. 1956. Self-incompatibility in rye. I. Genetic control in the diploid. Hereditas 42: 293-348.

LUNDQUIST, A. 1957. Self-incompatibility in rye. II. Genetic control in the tetraploid. Hereditas 43: 467-511.

LUNDQUIST, A. 1958a. Self-incompatibility in rye. III. Homozygosity for incompatibility factors in relation to viability and vegetative development. Hereditas 44: 174-188.

LUNDQUIST, A. 1958b. Self-incompatibility in rye. IV. Factors related to self-seeding. Hereditas 44: 193-256.

MAAN, S. S. 1973. Addition of rye chromosomes to wheat with rye cytoplasm. Proc. 13th Int. Congr. Genet., Berkeley, Calif. p. 5165.

MAAN, S. S., and LUCKEN, K. A. 1971. Male sterile wheat with rye cytoplasm. J. Hered. 62: 353-355.

MEDVEDEV, P. F. 1965. Gradual conversion of winter forms of rye into spring. [P.B.A. 40: 4873.] Nauch. Tr. Sev.-Zapad. Nauch.-Issled. Inst. Sel. Khoz. 9: 93-107.

MOLOTKOVSKII, G. H. 1950. Winter ryes with branding ears in Bukovina. [P.B.A. 21: 2599.] Selek. Semenovod. 9: 72-73.

MORI, T. 1948. Cited by Deodikar (1963).

MÜNTZING, A. 1943. Genetical effects of

duplicated fragment chromosomes in rye. Hereditas 29: 91-112.

MÜNTZING, A. 1945. Cytological studies on extra-fragment chromosomes in rye. III. The mechanism of non-disjunction at the pollen mitosis. Hereditas 31: 457-477.

MÜNTZING, A. 1951. Cytogenetic properties and practical value of tetraploid rye. Hereditas 37: 17-84.

MÜNTZING, A. 1957. Frequency of accessory chromosomes in rye strains from Iran and Korea. Hereditas 43: 682-685.

MÜNTZING, A. 1966. Accessory chromosomes. Bull. Bot. Soc. Bengal 20: 1-15.

MÜNTZING, A., and AKDIK, A. 1948. Cytological disturbances in the first inbred generation of rye. Hereditas 32: 473-494.

MÜNTZING, A., and LIMA DE FARIA, A. 1949. Pachytene analysis of standard fragments and large isofragments in rye. Hereditas 35: 253-268.

NAKAO, M. 1911. Cited by Gotoh (1931).

NECAS, J. 1961. The inheritance of branched ear in rye accompanied by increased number of flowers and flower plants. [P.B.A. 31: 3277.] Biol. Plant., Praha, 1961: 3: 65-74.

NĚMEC, B. 1910. Cited by Gotoh (1931).

NEUMANN, M. P., and PELSHENKE, P. F. 1954. Brotgetreide und Brot. (Bread cereals and bread). Parey: Berlin and Hamburg.

OINUMA, T. 1953. Karyomorphology of cereals. II. Karyotype alteration of rye Secale cereale L. Jap. J. Genet. 28: 28-34.

O'MARA, J. G. 1946. The substitution of a specific Secale cereale chromosome for a specific Triticum vulgare chromosome. Genetics 32: 99-100.

OSSENT, H. P. 1930. Perennial rye. Der Zuchter 2: 221-227.

OSTANIN, A. M. 1959. Developing winter or spring habit in biologically different forms of rye. [P.B.A. 30: 3853.] Agrobiologiya 3: 385-393.

PETERSON, R. F. 1934. Improvement of rye through inbreeding. Sci. Agr. 14: 651-668.

PURVIS, O. N. 1939. Studies in vernalization of cereals. IV. The inheritance of the spring and winter habit in hybrids of Petkus rye. Ann. Bot. N.S. 3: 719-730.

QUALSET, C. O., and HOSKINSON, P. E. 1966. A germplasm source population of rye. Crop Sci. 6: 219.

REES, H. 1955. Genotypic control of chromosome behaviour of rye. I. Inbred lines. Heredity 9: 93-116.

REES, H., and THOMPSON, J. B. 1956. Genotypic control of chromosome behaviour in rye. III. Chiasma frequency in homozygotes and heterozygotes. Heredity 10: 409-424.

RHOADES, M. M., and DEMPSEY, E. 1972. On the mechanism of chromatin loss induced by the B-chromosome of maize. Genetics 71: 73-96.

RILEY, R. 1955. The cytogenetics of the differences between some Secale species. J. Agr. Sci. 46: 377-383.

RILEY, R. 1965. Cytogenetics and plant breeding. Genetics today. Proc. 11th Int. Congr. Genet. 3: 681-688.

RILEY, R., and CHAPMAN, V. 1958. The production and phenotypes of wheat-rye chromosome addition lines. Heredity 12: 301-315.

RILEY, R., and EWART, J. A. D. 1970. The effect of individual rye chromosomes on the amino acid content of wheat grains. Genet. Res. Camb. 15: 209-219.

RILEY, R., and MACER, R. C. F. 1966. The chromosomal distribution of the genetic resistance of rye to wheat pathogens. Can. J. Genet. Cytol. 8: 640-654.

RIMPAU, J. 1973. Transmission und Meiose in alloplasmatischen Weizen-Roggen-Additionslinien. Z. Pflanzenzüchtg. 69: 275-300.

RIMPAU, J., BRANDES, D., and ROBBELEN, G. 1973. Cytoplasmic effects in wheat-rye addition lines. Proc. 4th Int. Wheat Genet. Symp., Columbia, Mo. p. 185-189.

RÜMKER, K. 1912. Neue Ergebnisse meiner Zuchtungsstudien auf dem Versuchfelde in Rosenthal. Z. Landwirt: 263-265.

RÜMKER, K., and LEIDNER, R. 1914. Ein Beitrag zur Frage der inzucht bei Roggen. Z. Pflanzenzüchtg. 2: 427-444.

SAKAMURA, T. 1918. Kurze mitteilung ueber die chromosomenzahlen und die Verwandtschaftsverhaltnisse der Triticum-arten. Bot. Mag. (Tokyo) 32: 151-154.

SÁNCHEZ-MONGE, E., and SOLER, C. 1973. Wheat and triticale with rye cytoplasm. Proc. 4th Int. Wheat Genet. Symp., Columbia, Mo. p. 387-390.

SARMA, N. P., and NATARAJAN, A. T. 1973. Identification of heterochromatic regions in the chromosomes of rye. Hereditas 74: 233-238.

SAX, K. 1935. Variation in chiasma frequencies in Secale, Vicia and Tradescantia. Cytologia 6: 289-293.

SCHIEMANN, E., and NURNBERG-KRÜGER, U. 1952. Neue untersuchungen an Secale africanum Stapf. II. Secale africanum und seine bastarde mit Secale montanum und Secale cereale. Naturwissenschaften 39: 136-137.

SCHNELL, F. W. 1960. Cited by Geiger and Schnell (1970).

SEARS, E. R. 1966. Nullisomic-tetrasomic combinations in hexaploid wheat. In: Chromosome manipulations and plant genetics, ed. by R. Riley and K. R. Lewis. Suppl. Heredity 20: 29-45.

SEARS, E. R. 1968. Relationships of chromosomes 2A, 2B and 2D with their homoeologous. Proc. 3rd Int. Wheat Genet. Symp., Aust. Acad. Sci., Canberra. p. 53-61.

SERSTOVA, K. N. 1959. An experiment on changing spring into winter rye in conditions of the Jakutian A.S.S.R. [P.B.A. 30: 3852.] Agrobiologiya 3: 378-384.

SHEPHERD, K. W. 1968. Chromosomal control of endosperm proteins in wheat and rye. Proc. 3rd Int. Wheat Genet. Symp., Aust. Acad. Sci., Canberra. p. 86.

SHMARGON, E. N. 1938. Analysis of the chromosome structure of mitotic chromosomes in rye. C. R. (Dokl.) Acad. Sci. U.S.S.R. 21: 259-261.

SPRAGUE, H. B. 1938. Breeding rye by continuous selection. J. Amer. Soc. Agron. 30: 287-293.

STEGLICH, L., and PIEPER, H. 1922. Vererbungs und Zuchtungsversuche mit Roggen. Fuhlings Landbw. Zbg. 71: 201-222.

STROUN, M., MATHON, C. C., and PUGNAT, C. 1960. The transformation of winter rye (*S. cereale*) into spring rye as a result of a modification of the conditions of vernalization. [P.B.A. 31: 3281.] Ber. Schweiz. Bot. Ges. 70: 440-447.

STUTZ, H. C. 1957. A cytogenetic analysis of the hybrid *Secale cereale* L. × *S. montanum* Guss. and its progeny. Genetics 42: 199-221.

STUTZ, H. C. 1958. A new macromutation in rye. Proc. Utah Acad. Sci. 34: 59-60.

STUTZ, H. C. 1972. On the origin of cultivated rye. Amer. J. Bot. 59: 59-70.

SYBENGA, J., and PRAKKEN, R. 1962. Gene analysis in rye. Genetica 33: 95-105.

SZENTIVÁNY, A. 1954. Forms and properties of branched ears of rye bred for branching. [P.B.A. 25: 2977.] Podohospodarstvo 1: 348-363.

THOMPSON, W. P. 1922. Lethal factors in cereals. West. Can. Soc. Agr. Proc. 3: 53-59.

TSCHERMAK, E. 1906. Uber Zuchtung neuer Getreidearten mit kunstlicher Kreuzung. II. Mitteilung Kreuzungstudien am Roggen. Z. Landwirt. Vers. Oesterr. I: 1-45.

TSCHERMAK, E. 1913. Ueber Seltene Getreide Bastarde. Beitr. Pflanzenzüchtg. 3:49-61.

VERMA, S. C., and REES, H. 1974. Giemsa staining and the distribution of heterochromatin in rye chromosomes. Heredity 32: 118-121.

WALTHER, F. 1959. Model experiments for the production of synthetic heterotic varieties in rye. Z. Pflanzenzüchtg. 42: 11-24.

WATKINS, R., and WHITE, W. J. 1964. The inheritance of anthocyanins in rye (*Secale cereale* L.). Can. J. Genet. Cytol. 6: 403-410.

WELLENSIEK, S. J. 1948. Breeding research on rye II. The direct influence of the pollen on grain weight. [P.B.A. 20: 257.] Landbouwk. Tijdschr. 60: 125-127.

WEXELSEN, H. 1969. Studies on the genetic basis of spring and winter forms in diploid and tetraploid rye. [P.B.A. 40: 4874.] Meld. Norg. Landbr. Hogsk. 48(24): 1-15.

WRICKE, G. 1969. Untersuchungen zur Verebung der selbstfertilitat beim Roggen. Theor. Appl. Genet. 39: 371-378.

CHAPTER 3

DISEASES, PESTS, AND PHYSIOLOGY OF RYE

STANISŁAW STARZYCKI
Plant Breeding and Acclimatization Institute
Radzików, Poland

PART ONE—DISEASES AND PESTS

I.   DISEASES CAUSED BY FUNGI

A.   Diseases Caused by Ascomycetes

POWDERY MILDEW—*Erysiphe graminis* DC

The powdery mildew fungus exhibits many specialized forms on particular genera of cereals and grasses. *E. graminis* f. *secale* infects species from the genus *Secale,* mainly rye (*Secale cereale* L.). The adaptation of various forms to species on which they develop is rather strict; thus, there is no danger of easy passage of *E. graminis* from one type of plant to others.

Occurrence. In comparison with other genera of cereals, rye is infected by *E. graminis* relatively rarely. At the end of spring just after heading, on leaves, leaf sheaths, stem, and occasionally on spikes, a greyish-white coating is present. This is the mycelium with spores. After a few days the coating thickens and becomes more or less feltlike. The spots become darker and attain a greyish-brown color. Cleistothecia with ascospores form and are detectable as black points. The fungus at first infects lower leaves and slowly migrates to the upper parts of the plant. Only in very favorable conditions can the fungus infect whole rye plants and lead to their premature death.

Symptoms. E. graminis is an exoparasite of rye. Mycelial hyphae which form a coating on leaves thrust haustoria into epidermal cells and in this way take up nutrients. On the mycelium, conidiophores form with colorless conidiospores (oidia), which are disseminated during the vegetative period of the plants. The conidia are elliptical, 24–30 μ by 12–15 μ. At a certain stage, dark, round, spherical fructifications (cleistothecia) are formed. Their diameter is 132–224 μ. This is the ascal stage of the fungus. The ascospores are colorless, and 20–23 μ by 10–13 μ in size. *E. graminis* is able to form asci on rye varieties cultivated in central Europe. The fructifications with asci winter on the remains of leaves and stems left after harvesting. In spring, the asci burst and the liberated spores infect developing plants. In the summer, plants are infected by *E. graminis* conidia.

Under favorable conditions *E. graminis* fructifications, after formation and a short period of maturation, may burst on the leaves. In this case ascospores infect plants similarly as conidia.

Control. Deep autumn plowing to destroy the mycelia and cleistothecia can be used for partial control. Excessive nitrogen fertilization should be avoided as this tends to weaken the resistance of plants. *E. graminis* mycelium can be destroyed by sulfur preparations. However, spraying of plants is economical only on particularly valuable plantations.

ERGOT OF RYE—*Claviceps purpurea* (Fr.) Tul.

During the period of maturation of cereals, especially of rye and more rarely of wheat and barley, some spikelets develop hard, slightly bent, blue-violet crescents with white interiors instead of normal grains. These are ergot bodies (sclerotia) of the fungus *C. purpurea.* The ergot bodies scatter during harvest or are collected with the grain. Flour prepared from grain contaminated with ergot is toxic to humans and animals.

Occurrence. Ergot is a common disease of rye. The form which occurs on rye may infect various genera of grasses *Dactylis glomerata, Festuca pratensis, Briza media, Arrhenatherum elatium,* and some species of the genus *Poa.* Biological forms of *C. purpurea* which occur on *Poa annua* and brome grass infect only these species.

Infection of rye by ergot is favored by humid summers and sparse planting. Most frequently infected plants are the marginal and highest ones as these are visited more frequently by insects. Infection of rye by ergot occurs readily because rye blooms with open florets.

Symptoms. The first symptom of infection is the appearance of drops of sweet liquid on young spikes. This liquid contains colorless elliptical conidia, 3.5 μ by 7 μ. Insects feed on this liquid and subsequently transfer conidia to the spikes they visit. The mycelium develops in the spikelets instead of the caryopsis. The dimensions of the sclerotia formed during the development of rye are 3–5 mm by 1–5 mm. The size depends on the number of infected flowers in the spikelet. Some sclerotia fall to the ground during harvest. The fungus over-winters in the form of sclerotia which germinate in the spring, forming stalked structures terminated by a small (diameter, 1–2 mm) spherical head. There are numerous dark points on the head. These are the outlets of the perithecia. The perithecia are bottleshaped with dimensions of 200–250 μ by 140–170 μ. The asci, which form in the perithecia, are cylindrical. The ascospores are filamentous, measuring 50–78 μ by 1 μ. The length of the stalked structures depends on the thickness of the soil layer with which they are covered during plowing. If the sclerotia, which scatter during the harvest, are covered with a soil layer 10 cm thick, the germinating sclerotia do not emerge from the surface.

Control. Control of ergot is not complicated and is mainly based on the use of appropriate agrotechnical practices. Careful performance of the first plowing and use of deep plowing decrease the danger of development of the parasite in the soil. Rye seed should be carefully cleaned of sclerotia. So far, attempts have not been made to obtain rye varieties resistant to ergot.

Uses of ergot sclerotia. Sclerotia are used as the source of alkaloids for medicinal purposes. In many European countries, sclerotia are collected. As the amount under normal cropping is not sufficient, ergot is collected from rye fields artificially infected with spores. Kumelowska (1961) evaluated several rye varieties with respect to their suitability for ergot growth. She found that some varieties are more sensitive to infection. Thus, varieties resistant to some extent do exist and it may be possible to select these by plant breeding.

TAKE-ALL OF WHEAT—*Gaeumannomyces graminis* Arx et Olivier — *Ophiobolus graminis* Sacc.

Occurrence. The mycelium of this fungus may survive saprophytically in the soil for several years. *Ophiobolus* is sensitive to lack of oxygen and excess carbon dioxide in the soil. For these reasons it develops on rye, especially when it is grown on sandy soils. Besides rye, the parasite infects wheat, barley, soybean, corn and some grasses.

Symptoms. Soon after heading of rye on the still-green field, smaller yellow or white plants with underdeveloped spikes without grain or with undeveloped grain may be observed. Such plants can readily be extracted from the soil. The base of the tillers of diseased plants is black and covered by a brown-grey coating composed of hyphae. The mycelium of the parasite enters the root tissue through the root neck. Dark mycelium may be observed within the leaf sheaths of lower leaves. Perithecia form on dying tillers. They are dark, pear-shaped, and 350–500 μ in size. In the perithecia, elongated asci form, containing multicellular ascospores, 13–15 μ long and 2–5 μ wide. The fungus winters in the form of perithecia with ascospores.

Control. Fields in which the disease has occurred to a considerable extent should be excluded from production for 2–3 years. Deep plowing should be used before planting. Beets and legumes should be grown as a forecrop as they considerably decrease the occurrence of the disease. So far, there has been no serious plant breeding to develop varieties that are resistant to this fungus.

B. Diseases Caused by Basidiomycetes

STALK SMUT—*Urocystis occulta* (Wallr.) Rab.

The disease stalk smut is caused by the fungus *Urocystis occulta* (Wallr.) Rab. It is prevalent in central Europe and infects mainly rye.

Occurrence. The symptoms of stalk smut are usually found in neglected rye fields and on soils with poor structure. Natural pH of the soil and dry conditions favor the development of the fungus. It does not cause serious economic losses.

Symptoms. During flowering, long, lead-grey or black streaks appear on stems, leaves, leaf sheaths and sometimes also on the spikes. Later, these streaks become more distinct and the skin covering them breaks, liberating a mass of black spores. As the development of the fungus proceeds, the leaves separate into longitudinal bands. The stem usually breaks above the uppermost node and the empty spike twists and breaks.

In contrast to other types of smut, stalk smut mycelium develops mainly in the stem tissues and forms spores (chlamydospores) in groups of several cells, of which only some can germinate and the rest are filled with air.

Spores winter in the soil or on the surface of healthy rye. During germination of the grain, infectious hyphae easily penetrate the tissues of young shoots. The greatest intensity of chlamydospore formation may be observed during or just after flowering.

Control. The prevention of stalk smut is based mainly on the application of chemical agents for disinfecting the seed before planting. Rational agrotechnics such as proper dates of planting and fertilization with phosphorus and potassium also help to control this disease.

STEM RUST—*Puccinia graminis* Pers., f. *secalis*

Stem rust is one of the most dangerous diseases of cereals and grasses. This fungus has many forms, adapted to various members of the Gramineae family. The form *secalis,* which occurs on rye, barley, brome grass, meadow grass and orchard grass, is among the most common. Within each form, a number of races or biotypes may be distinguished which, in general, form by crossing in spermogonia on the intermediate host barberry or by mutations and somatic hybridization. Mains (1926) described 14 physiological races of *P. gr. secalis* found on various varieties of American rye.

Occurrence. *P. gr.* f. *secalis* has a full developmental cycle on two hosts. The stage of uredospores and teliospores is formed on cereal tissues and on wild and cultivated grasses while the aecia and spermogonia (pycnia) stages occur on the leaves of the barberry (*Berberis vulgaris*) and, in rare cases, on *Mahonia aquifolium.*

The uredospore stage on rye stems appears usually toward the end of June (in the northern hemisphere) and the teliospore stage at the end of the vegetative period. The aecia appear at the end of May or in June on the lower part of barberry leaves in the form of orange clusters. Aeciospores are spherical, 14–16 μ in diameter. The pycnia, from which pycniospores are liberated, form on the upper surface of the barberry leaf.

In the climate of central Europe, *P. gr.* f. *secalis* winters in the form of teliospores which survive in the low temperatures. Wintering of uredospores is rare in this climate. In warm climate regions uredospores are the main source of infection. The source of primary infection in central Europe are the basidiospores produced by the teliospores. Detailed investigations have shown that the extent of infection of plants by *P. gr.* f. *secalis* decreases with the distance of barberry plants from the cereal growing area.

The development of the fungus on infected plants is favored by moderate temperatures and high humidity. It has been found that temperatures in the range 12°–23°C and 100% relative humidity favor germination of uredospores while high humidity and temperatures of 20°–22°C favor infection.

Symptoms. Elongated rust brown pustules form on the leaf sheaths, stems, sometimes on the leaves, and very rarely on spikes. Uredospores with dimensions of 17–40 μ by 13–23 μ are unicellular, ellipsoid, and covered by sparse spikes. Telia form at the end of the vegetative period near the uredia. Dark brown or, more frequently, black telia contain teliospores which measure 35–60 μ by 12–22 μ and are two-celled, spindle- or club-shaped, rounded or tapering at top, and set on a strong peduncle.

P. gr. f. *secalis* developing on barberry and Mahonia causes formation of orange or red spots on the leaves. On these spots, on the upper side of leaves, spermogonia develop under the epidermis. Orange aecia form on the opposite side of the leaves.

Control. The most common means of control is by agrotechnical procedures which limit the growth of *P. gr. secalis.* To destroy the teliospores, deep plowing should be used after the first regular plowing. Early planting causes a quicker development of the plants which makes infection by aeciospores more difficult or impossible. Excessive nitrogen fertilization should be avoided. If possible, grasses which are hosts of the fungus should be destroyed. It is important to remove barberry shrubs growing in the vicinity of rye growing areas.

BROWN LEAF RUST—*Puccinia dispersa* Erikss.

Leaf rust is a very common parasite and is among the most frequently found fungi on rye leaves. This disease causes premature drying up of the leaves which adversely affects the quality of the crop, grain yield and quality, and feeding value of the straw. In the climate of central Europe the most extensive development of leaf rust occurs during the period of full development of the host.

Occurrence. *P. dispersa* is a two-host rust with a full developmental cycle. Uredospores and teliospores form on rye, while aecia and spermogonia form on the leaves of plants from the family Boraginaceae, *Lycopsis arvensis* and *Anchuza officinalis*.

The development of one generation of uredospores may take from 6–10 days, depending on the temperature. Temperatures of 20°–30° C are optimal. The teliospores can germinate in the autumn. The basidiospores infect the leaves of the intermediate host, on which many clusters of aecia form and are disseminated by the wind, and then infect winter crops. The intermediate host is not necessary for the development of *P. dispersa*. It may winter in the form of mycelium and uredospores on self-sown cereal plants and winter crops. Uredospores are resistant to low temperature and do not lose the ability to germinate at −26° C if they are covered with snow.

It has been observed that varieties of rye planted later during the normal seeding period are infected by *P. dispersa* to a lesser extent. *P. dispersa* is not a biologically homogeneous species. Many physiological races adapted to different varieties occur.

Symptoms. The first symptoms of infection are observed in rye at the beginning of May. Rust brown uredospore pustules appear on the upper part of leaf blades, occasionally on the lower side of the leaves, and on the leaf sheaths. Uredospores are unicellular, spherical or elliptical, and covered with spines, 22–26 μ by 22–31 μ. During the period of rye maturation, black oval pustules of teliospores are formed mainly on the lower side of the leaves. These spores consist of two cells, which are elongated or club-shaped, on pedicels. The spore dimensions are 40–57 μ by 12–22 μ. In strong infection the spores cover the whole surface of the leaf blades, preventing assimilation.

Control. Some control can be achieved by adhering to proper planting times. Destruction of post-harvest remains and weeds from the family Boraginaceae offers an additional measure of control.

YELLOW RUST—*Puccinia striformis* West—*Puccinia glumarum* Erikss. et Henn.

Yellow rust occurs in central Europe on rye only under exceptional conditions. It occurs more commonly on grasses. Gäumann (1959) cites 80 species of grasses from 21 genera as its hosts. Quantitative and qualitative losses caused by yellow rust infection are due to a decrease of chlorophyll content and an increase in transpiration.

American investigations have shown that germination of seeds from strongly infected plants may decrease by 50% (Kochman, 1973). Seeds from infected plants dry up earlier during development and usually have a lower protein content.

The development of the disease and infection are favored by low spring and early summer temperatures and longer periods of humid weather. Infection during flowering causes a considerable decrease in the weight of roots and grain.

Symptoms. The disease can be detected by the appearance of pustules of yellow-orange elongated uredospores (0.5–1.0 mm by 0.3–0.4 mm) on leaf blades and sheaths and on stalks in June and July. After infection, symptoms of the disease may occur on husks, glumes, awns and young caryopses. This is accompanied by severe leaf chlorosis. Small dark brown or black spores (teliospores) are formed during plant maturation on the lower side of the leaf blade, in the leaf sheath, and on the stem.

It is believed that *P. striiformis* does not have a complete cycle. The uredospores are spherical or broadly elliptical with dimensions of 14–26 μ by 12–23 μ, with a rather thin cell wall, covered by papillae. Teliospores are club-shaped and frequently asymmetrically placed on short stalks rounded at top with two outgrowths. Teliospores can germinate during the same summer but may also germinate in the spring. In the climatic conditions of central Europe, this parasite winters in the form of mycelium or uredospores on the leaves of self-sown grain, winter crops or perennial grasses.

Control. Correct agrotechnical practices such as deep autumn plowing help to control the disease. Application of appropriate amounts of P and K fertilizers is helpful. Excessive amounts of N should be avoided. Crop rotations with a minimum three-year interval between rye plantings have been successful in decreasing the level of infection. Self-sown plants and wild grasses, which are a source of yellow rust uredospores in the early spring, should be destroyed.

RESISTANT VARIETIES OF RYE

Puccinia dispersa Erikss. Results of investigations from the 1920's indicate that resistant varieties can be bred (Mains, 1926; Garbowski, 1928). This possibility has been confirmed by Garbini (1950). In Florida, inbred lines of rye resistant to *P. dispersa* have been obtained. U.S. rye varieties, Weser and Emory, are resistant to this type of rust.

Puccinia graminis Pers. f. *secalis.* In Poland, Czarnocka (1939) observed that the rye variety Puławskie Wczesne is resistant to *P. gr. secalis.* A certain degree of resistance is shown by the Swiss rye variety Rothenbinnen. In Austria, the varieties Crysanth Hauser and Lurnfield are tolerant to *P. gr. secalis.* In years of epiphytotics, the variety Lurnfield gave yields higher by 25-45% than other varieties grown.

Sources of genetic resistance to other species and genera are being sought. The possibility of transferring resistance from wheat to rye was investigated by Waterhouse (1953). Biotypes of *S. montanum* Guss. resistant to *P. gr. secalis* have been identified (Anonymous, 1953). The results of other research workers also indicate that breeding of resistant rye varieties is possible and that selection for resistance increased the yield and resistance to lodging.

C.   Diseases Caused by Deuteromycetes (Fungi Imperfecti)

EYESPOT (Foot Rot or Culm Rot)—*Cercosporella herpotrichoides* Fron.

Eyespot is now considered as one of the most dangerous diseases of cereals. The causal organism *Cercosporella herpotrichoides* was isolated for the first time by Fron in 1914 in France. This fungus attacks various types of cereals (Ecochard, 1963). However, the genus *Secale* appears to be more resistant to the fungus than the genus *Triticum*.

Occurrence. Between the two wars and during World War II numerous communications appeared on the occurrence of the disease in France, United States, England, Holland, Germany and Poland. During this period the biology and ecology of the pathogen was investigated and preventative measures were reported. The intensity with which *C. herpotrichoides* occurs in various areas where cereals are cultivated differs. The fungus has been observed on cereals grown on both light and heavy soils. In conditions favorable to fungal development, after the plants are infected, the losses in yield may be considerable. The investigations of Lechner (1961) showed that yields of the rye varieties Tylstrup I and Tylstrup II, planted in a locality where grain had been cultivated for several years, were decreased by 56.7% and 31.8%, respectively. The extensive decrease in yield was attributed to the diseases caused by *C. herpotrichoides* and *Ophiobolus*.

Several authors have stated that the disease that causes breaking of stems is spreading dangerously in western and central Europe. The immediate cause of the intensification of the disease is the increase of cereal grain cultivation. Succession of cereals, especially of wheat, rye, and barley, in crop rotation causes accumulation of mycelium in the soil. The development of mycelium in western and central Europe is favored by a high water content in the soil and air.

Symptoms. The first symptoms of the disease on plants may be observed in late autumn. Light brown spreading spots are formed on the leaves of rye. Whole leaf blades may become yellow or brown. Typical symptoms in the form of oval spots with dark margins form on the lower parts of the leaf blade in the region of the first node during heading of the plants. The oval shape of the spots is characteristic of the disease. During this period the inside of the stems is overgrown with the grey mycelium of the parasite. Black bodies of the fungus (sclerotia) form on the surface of the spots. The conidia are wedge-shaped. They consist of 7–8 cells and are 50–70 μ long and 2–3 μ wide at the base. Each cell has a more or less distinct structure. Its depth depends on the weather conditions, mainly the relative humidity. The conidia, in general, form on the stubble after the rye is harvested. In the same year, conidia may infect new cereal plants. During warm and humid autumns conidia may again be formed. Under such conditions the fungus spreads quickly.

Control. So far, few experiments on the control of eyespot in rye have been carried out. Most of the investigations concerned wheat. Species of the genus *Secale* are less sensitive to *C. herpotrichoides* than species of the genus *Triticum*. In general, it is considered that planting cereals one after another increases the danger of occurrence of the disease.

Manuring affects the degree of infection of cereals by *C. herpotrichoides*. It has been shown that application of K and P fertilizers does not affect the occurrence of eyespot. Application of N fertilizers favors the development of the fungus, especially if high rates are used. Also, the time of N fertilizer application seems to affect the degree of infection. It is considered that top-dressing with N during the period from shooting to heading is better than earlier additions.

The seeding density may also be of considerable importance. High density planting increases the danger of infection.

After the harvest, plant debris, self-sown cereals, and some wild-growing grasses may pass the disease to new plants. Therefore, careful removal of the weeds from the fields and early plowing, with frequent harrowing of fields during the period preceding deep plowing, limit the expansion of the parasite to a considerable extent.

Breeding of resistant varieties. There has been little work on breeding rye varieties that are resistant to eyespot. The reports of Lange-de la Camp (1966) and Lechner (1961) indicate that it may be possible to breed rye varieties with a decreased sensitivity. More detailed investigations were carried out by Bojarczuk (1968) and Bojarczuk and Bojarczuk (1972). Testing of the degree of infection of the plants by the parasite was performed by the method elaborated by Macer and Van den Driessche (1966). The infection of several dozen rye varieties and breeding lines by eyespot under conditions of strong infection was rather high, 62% of the plants on the average. Varieties and lines differed from each other significantly in respect to resistance. The German variety Carsten and the Polish varieties Smolickie and Dańkowskie Selekcyjne exhibited a high degree of tolerance.

SNOW MOLD—*Fusarium nivale* (Fr.) Ces.

The disease called snow mold occurs in winter grain crops. It is most frequently observed on rye. Investigations performed in many agricultural regions indicate that the disease may be caused by various fungi but most frequently by *Griphosphera nivalis* (Schaffnit) Müller et v. The conidial stage of this fungus is known as *Fusarium nivale* (Fr.) Ces. (Railo, 1950; Jamalainen, 1959; Germ, 1960; Benada, 1963). Other species from the *Fusarium* genus: *F. avenaceum, F. culmorum, F. antiquoides, F. poae, F. scirpi,* and *F. sporotrichoides* may cause snow mold (Łacicowa and Filipowicz, 1972). The occurrence of these species on rye is rather sporadic even though Atanasoff (1923) ascribed a considerable role to *F. avenaceum* in rye infection in Germany. Pielka (1968) isolated *F. avenaceum* from rye plants infected with snow mold in Poland and confirmed experimentally the strong pathogenic action of this species. In eastern regions of Europe (U.S.S.R.) snow mold may be caused by *F. graminearum* Schw. (Tyunov *et al.*, 1969).

Occurrence. Symptoms of snow mold occur with various intensities where rye, especially winter rye, is cultivated. The intensity of snow mold occurrence in particular years depends on the weather. According to Pichler (1952), the rye-growing areas which run the greatest risk of snow mold are those on which the snow cover persists for 100 days of the year. Other authors suggest that the fungus destroys plants remaining under the snow for 80 days, especially if the soil is not strongly frozen under the snow (Stachyra, 1959). The regions with the highest snow mold risk include Scandinavia and the mountain regions of central Europe.

Symptoms. Infected seeds or mycelium growing saprophytically in the soil may be the source of infection. Seeds, especially weakly infected ones, may germinate in the soil, but the plants which develop from them are weak. Dead leaves are often seen in young infected plants. The leaves twist during passage through the soil and often die before reaching the surface. Rye infected by the mycelium developing in the soil generally does not show definite symptoms of the disease in the autumn. Fusariosis of seedlings develops in early spring, especially in plants in unfrozen soil covered by a thick snow layer. In early spring, after the snow melts, smaller or bigger patches of dead or dying plants may be observed. The plants are covered by an abundant white or slightly pink mycelial coating. The change of the white mycelial color to pinkish white is due to the formation of numerous conidia in the sporodochia on the surface of dead leaves. One- or two-celled conidia have a crescent shape and dimensions of 9–23 μ by 2–4 μ. The ascus stage (*Calonectria graminicola*—Berk. et Brme.) may form on dry tissues of infected stems, leaves or leaf sheaths. The fungus winters in the soil or on the remains of dead grain tissues in the form of mycelium or peridia of the ascal stage. Mass infection of young plants occurs in early spring.

FUSARIOSIS OF RYE SPIKES

The symptoms of fusariosis of rye spikes are caused by the development of several *Fusarium* species: *F. avenaceum* Fr. (Sacc.), *F. nivale* (Fr.) Ces., *F. culmorum* (W.G.Sm.) Sacc., and *F. graminearum* Schw. (Table I).

Fusarioses of spikes may be recognized by pink or reddish coating on spikelets. The main consequence of infection is poor grain formation. Often the infected spikes are white and their caryopses are dry and not fully developed. Sometimes the spikelets are empty. Fusarioses of spikes are occasionally accompanied by the darkening of the base of the stem.

Fungi-causing fusarioses live in almost every soil in regions where rye is cultivated. The development of fungi in plant tissues takes place first of all under conditions that cause a weakening of natural plant resistance. Excessive moisture and acidity of the soil contribute to the development of the mycelium and to the weakening of the plants. Infected plants are usually seen in fields in which crops have not been properly rotated or correctly cultivated.

Control. Fusarioses can be controlled by careful cultivation of the soil, using deep autumn plowing and appropriate application of K and P fertilizers. Proper

TABLE I
Size and shape of conidia

Species of *Fusarium*	Conidia	
	Shape	Size
		μ
F. nivale	crescent shape, one or two cells	2–23 $\times$ 2–4
F. avenaceum	four cells	36–78 $\times$ 3–5
	five cells	37–70 $\times$ 3–4
	six cells	38–78 $\times$ 3–5
F. culmorum	fusiform crescent, five cells	29–41 $\times$ 6–8
	six cells	24–47 $\times$ 6–7
F. graminearum	fusiform cresent with a distinct foot	...

dates for sowing and harvesting should be observed and the seeds should be treated with mercury preparations before planting. Tyunov *et al.* (1969) recommended that infected grain should be heat treated at 80° C before it is fed to animals.

Breeding of resistant varieties. The losses of winter rye caused by the infection of fields by *F. nivale* are extremely large. Accordingly, there have been many attempts in Europe to breed varieties that are resistant to this disease. The resistance of plants to Fusarium may result from a number of different factors (Roemer *et al.*, 1938; Klinkowski *et al.*, 1965). It may be due to peculiar anatomy and morphology of plants or to their biological and physiological characteristics. A detailed laboratory method for the detection of resistance of rye seedlings and plants to *F. nivale* and *F. culmorum* was elaborated by Baltzer (1930). Pichler (1952) proposed and elaborated methods of plant infection under natural growth conditions.

So far, no plants that are fully resistant to *F. nivale* or other *Fusarium* species have been obtained. Cultivated varieties of rye differ in degree of susceptibility to this disease but the differences are minor. In regions where rye has been cultivated for a long time, ecotypes resistant to snow mold to a greater degree than varieties introduced from other regions have been selected. Brundza (1943) found that in Lithuania, local varieties of rye are more resistant to snow mold than imported German and Swedish varieties.

Velikovskii (1974) initiated the development of resistant varieties. He tested the resistance of many varieties to snow mold and found that the varieties Ensi (Finland), Burańskie (Czechoslovakia), Lunganer Taueru (Austria), and Vyatka (U.S.S.R.) have some degree of resistance. On the basis of this work, he proposed a method for breeding for resistance, assuming that the resistance is due to one gene.

Kobylyanskii and Ilichev (1972) reported that the following varieties are resistant to eyespot: U.S.S.R. varieties Vyatka 2 and Kaukaskaya, Finnish varieties Jo 090, Kahozel, Pekka and Onni, Austrian varieties Stalker Lassaer, Kefermartker and Edelhofer, Polish varieties Ottarzewskie and Kazimierskie, Czechoslovakian variety Doupooske, and German varieties Bergroggen and Lichover.

SPOT BLOTCH—*Helminthosporium sativum* (Pam) King et Bakke

Spot blotch is caused by the fungus *Helminthosporium sativum* (Pam) King et Bakke [*Bipolaris sorokiniana* (Sacc. in Sorok) Shoemaker]. The parasite has been noted on rye and much more frequently on barley and wheat.

Occurrence. The disease is mainly prevalent in North America. It has also been noted in the U.S.S.R. but is of little economic importance in Europe.

Symptoms. The first symptoms appear on seedlings. The leaves of young plants are covered with a dark brown coating of mycelium with conidia. The leaves lose vigor and droop to the ground. On older leaves, elongated or round spots appear. These spots are dark brown with a yellow margin. The borders of the spots are not distinct. As a result of the development of the parasite, the leaves dry out and fall. Plant development ceases prematurely without formation of seeds. The fungus also attacks spikes, causing formation of brown spots on glumes and caryopses. Infected caryopses do not germinate. After moistening of the infected caryopses, abundant mycelium develops on their surface. The pathogen may spread from infected seed to the underground part of the plant, the base of the stalk and roots.

Conidia of the fungus appear only on dead fragments of tissues. The conidia are spindle-shaped, sometimes slightly twisted, with dimensions of 60–120 μ by 15–20 μ. They germinate by two vegetative hyphae.

Control. The frequency of cereal cultivation should be decreased on fields infected with *H. sativum*. The cultivation of quickly growing varieties is also recommended. Positive results are also obtained by treatment of seeds with appropriate chemical fungicides.

BLACK MOLD—*Cladosporium herbarum* (Pers.)

Black mold frequently occurs on plants weakened by infection with *Ophiobolus graminis* or *Cercosporella herpotrichoides*. This fungus does not cause extensive economic losses. In exceptionally humid years, it may lead to the significant lowering of grain quality.

Occurrence. The fungus spreads as conidia. A black coating of conidia forms on the surface of grains, stems, husks, glumes, and leaves. Their stalks are not branched. They grow in bunches and are olive in color. Conidia form in chains of 2–3. They are egg-shaped, 10–15 μ by 4–7 μ, and consist of 3–4 cells.

Symptoms. In general, a black coating of the fungus on spikes and stems may be observed during the period of maturation in warm and humid weather. In heavy infection, the fungus can cover all of the stem and spike.

Control. Deep plowing before the winter and early first plowing in the spring are the most successful means of controlling black mold. Rye harvested during rainy weather should be immediately dried to prevent development of the mold.

ANTRACNOSE—*Colletotrichum graminicola* (Ces.) Wils.

Antracnose causes increasing losses of the rye crop. Frequent changes of weather favor development of the disease. Symptoms are mainly visible in the lower parts of plants. The first symptoms are rusty brown spots on young seedlings. The surface of the spots is covered with numerous stalks of conidia. The conidia are unicellular, 8–15 μ by 4–12 μ. The fungus also infects rye stems. The infected plants are more sensitive to the attack of other fungi. The fungus spreads most quickly at 28°C.

Control. Rational soil cultivation with deep plowing and appropriate crop rotation are the most effective means of controlling antracnose.

SEPTORIA LEAF BLOTCH—*Septoria secalis* (Rob.)

Septoria leaf blotch infects only rye. It does not cause serious economic losses in any of the rye-growing areas.

Occurrence. During the vegetative period of the rye plant, the fungus disseminates by means of conidia formed in numerous pycnidia. The conidia are colorless, straight or slightly bent, 28–40 μ by 2–4 μ, with three septa. The conidia leave the pycnidia most easily in humid weather and germinate at 20°C. The parasite also forms rod-like microspores, 10 μ by 0.5 μ. The ascal stage winters on remains of plants and the mycelium on infected grain.

Symptoms. Light brown spindle-shaped spots (about 2 cm long with a straw-yellow center and dark margin) appear on leaves and sometimes on leaf sheaths. Often the shape of the spots is irregular, especially when they occupy the whole leaf blade. Dark pycnidia, visible with the naked eye, form on the surface of the spots. The lower leaves of plants are infected most. Sometimes the symptoms may occur on husks, causing a dark coloring. During heavy infection, Septoria may destroy up to 50% of plant shoots.

Control. Chemical treatment of the seed is the best method for controlling Septoria leaf blotch. Rational cultural practices, correct crop rotation with at least 1-year intervals between rye crops, and destruction of weeds and after-harvest crop remains on which the fungus may winter help to lessen the degree of infection.

LEAF BLOTCH OF RYE—*Rhynchosporium secalis* (Oudem.) I. I. Davies

Leaf blotch of rye is common mainly in the colder rye-growing regions. In Europe, it has been found in regions adjacent to mountains. It has not caused significant economic losses in central Europe. Serious crop losses have occurred from this disease in the United States; infection of 75–100% of the plants can reduce grain yield by about 8%.

Occurrence. Infection is by means of conidia transported over large distances by wind and rain. Conidia form on the surface of leaf blades. They are 2-celled, cylindrical, and 12–20 μ by 2–4 μ in size. At lower temperatures, the parasite produces more conidia and shows greater activity. The mycelium and conidia winter in the remains of grains or weeds.

Symptoms. The disease symptoms are visible in early spring as dark grey-blue spots on leaf blades and occasionally on leaf sheaths. The center of the spots gradually becomes white or light brown while the periphery is brown or yellow. In severe infection, the leaves usually dry up. Intensification of the symptoms takes place during heading.

Control. The best means of controlling rye leaf blotch is by rational agrotechnical management. Deep plowing, destruction of post-harvest crop remains, and 3–4 year crop rotations have been found helpful for decreasing the incidence of the disease.

II. DISEASES CAUSED BY VIRUSES

A. Barley Yellow Dwarf

Barley yellow dwarf disease is caused by the virus *Hordeum virus nanescens.* The development of symptoms of this disease on rye has not been observed in Europe. The virus is transmitted by aphids, mainly of the species *Rhopalosiphum maidis* (L), *Sitobium avenae* (F), and *Macrosiphum dirhodum* (Walker). It is not transmitted by seed and soil. Infected plants grow more slowly, tiller more intensively, and form abnormal spikes. The normal development of caryopses is disturbed. The leaves of older plants become yellow at the tip and sides.

The most effective means of controlling barley yellow dwarf is by destroying the aphids that transmit the disease.

B. Wheat Dwarf

In Europe, the infection of rye by wheat dwarf virus is not heavy. The growth of the infected plants is arrested and they head poorly. Initially chlorotic yellowish spots form on the leaves and then extend to the whole leaf blade in later stages of the disease. The diseased plant dies quickly. In diseased cells, degeneration of chloroplasts takes place together with the appearance of large vacuoles. The nucleus, together with nucleoli, is enlarged but does not contain any protein crystals. The virus is transmitted by insects, jerboas from the species *Psammotettix alienus* Dahlb. and *Psammotettix striatus* L., and by the aphid *Rhopalosiphum oxycanthae* Schrk.

Infection by wheat dwarf virus can be reduced by control of leaf-hoppers and aphids and by application of rational agronomic practices.

C. Soil-Borne Mosaic

Soil-borne mosaic disease is caused by the virus *Marmor tritici*. Its occurrence was observed for the first time in 1919 in Madison County, Illinois, U.S.A. Two strains of the virus are known; one causes yellow, the other green mosaic of the leaves.

The disease symptoms on rye can be seen in early spring in the form of streaks and spots on leaves and sometimes also on leaf sheaths and husks. Affected plants are dwarfed and numerous undeveloped shoots appear. Plants attacked by green mosaic form an abundant bluish-green leaf rosette. Diseased plants also exhibit symptoms of physiological diseases in spite of appropriate fertilization, symptoms of frost bite, and others. The virus may be transmitted by means of soil and may persist in heavy soils for several years. Soil-borne mosaic is not transferred by the seeds. Plants weakened by unfavorable climatic conditions are infected easily. There is no effective method of controlling this disease in rye.

D. Oat Blue Dwarf

The oat blue dwarf virus attacks rye to a small extent. Infected plants grow poorly and tiller more strongly. The grain is underdeveloped. Light green spots which turn yellow and brown appear on leaves. The leaves eventually die. The virus vectors are the leaf-hoppers *Calligypona pellucida* F., *Calligypona discolor* Boh., and *Dicranotropis hamata* Boh., which transmit the virus mechanically. The virus is not transmitted by the soil or with seeds. Wintering of the virus in the field remains of infected plants is possible.

III. DISEASES CAUSED BY NEMATODES

A. *Ditylenchus dipsaci* Kühn

The nematode *Ditylenchus dipsaci* has been found on all continents. It is biologically heterogeneous and about a dozen races have been distinguished. *D. dipsaci* larvae, which winter in the soil and on plant remains, attack rye plants. In spring, the nematodes leave the plant remains and enter the root system and from there they migrate into the stem and leaves.

The symptoms of infection appear mainly in the spring and sometimes during the autumn, especially if the weather is unusually warm. The diseased plants are dwarfed, have dark leaves, tiller strongly, and form short thickened stalks. The leaves die prematurely, and the plant frequently dies before the formation of stalks or does not pass into the heading stage. In affected plants, the root system is poorly developed. The grain yield of the infected plants is very low. Winter rye is infected more strongly than spring rye. During maturation most of the nematodes emerge from the plants and enter the soil. Some remain and winter in the straw. The race of nematodes on rye is not strictly specialized. *D. dipsaci* can cause serious economic losses.

D. dipsaci is difficult to control as it occurs in many species of plants. Use of appropriate rotation of crops is recommended; maize, clover (both white and Swedish), rape, hemp, pea, field bean and forage beets should be avoided as forecrops. Carrots, wheat, potatoes, turnips, lupin, serradella, sainfoin, alfalfa and white mustard are good forecrops. In the infected fields, remains should be collected, removed, and destroyed after harvest. Heavy soils should be drained of excess water. Heavy applications of K and N fertilizers have been useful in controlling this pest.

B. *Anguina tritici* (Steinbuch)

Rye is infected by *Anguina tritici* to a lesser extent than wheat. Symptoms of infection may occur in autumn on young rye seedlings. Wrinkling and rolling of leaf blade edges take place. In adult plants the leaf often cannot emerge from the leaf sheath; it twists during growth and is deformed. The spike may be similarly deformed. Diseased plants are shorter than healthy ones.

Spikes that are strongly infected by the nematode become yellow. If the spike develops, galls of the parasite containing thousands of nematodes replace the grain. These galls are light yellow. The nematode occurs in nearly all countries in which wheat is grown.

The use of clean healthy seed and appropriate crop rotation helps to control this pest. In rotating crops, rye should not follow wheat. Farm manure from straw of infected plants should not be used as fertilizer.

C. *Heterodera avenae*

Rye is infected by *Heterodera avenae* only sporadically. The first symptoms of infection are seen on seedlings. Plants become stiff and, at the stage of the fourth leaf, turn brown. Eventually the leaves die. The growth and development of diseased plants are slow, and tillering is weak or inhibited by intensification of the disease. Parasitized roots of plants become dark and die. At the sites of feeding of *H. avenae,* thickenings develop which in time are torn open and *H. avenae* females emerge from them. Parasite development is favored by light soil. Stronger summer rainfall which speeds up plant development decreases the losses caused by *H. avenae.*

The most effective control of *H. avenae* is appropriate crop rotation. In soils infected with the nematode, leguminous and root crops should be planted. Early spring planting helps to decrease the losses from this pest. Green fertilizers reduce the *H. avenae* population in the soil.

IV. OTHER RYE PESTS

Rye may be destroyed by many other pests. In respect to their economic significance, three groups may be distinguished. Following is a more detailed description of some of the most harmful species of each group.

A. Pests Which Cause Major Damage

PESTS CAUSING CONSIDERABLE DAMAGE

The European earwig, *Forficula auricularia* L.; the leaf beetle, *Aelia acuminata* L. (family Dipodidae); carrion beetles, family Silphidae; cereal leaf beetle, *Oelema melanopus* (L.); gall midges, families Cecidomyidae and Itonididae; leaf-miner flies, family Agromizidae; and nematodes present a major problem.

Leaf Beetle—Aelia acuminata L. Occasionally, *Aelia acuminata* L. occurs in large numbers on rye. The insects suck up the juices from the leaves and spikes, causing the formation of white spots on the leaves and underdevelopment and whitening of the spikelets. *Aelia acuminata* produces one generation per year. It winters as the adult insect. The larvae feed on leaves, stems, spikes and young grain. Early infection of spikes and grain leads to sterility or underdevelopment of the seed. In central Europe, infection of up to 50% of spikes in rye crops has been observed.

Early planting of spring cereals, the use of fertilizers for accelerating the growth, early first plowing, and the use of chemical agents in case of mass infection are the recommended means of control.

B. Pests Which Cause Moderate Damage

PESTS THAT FEED ON RYE AND CAUSE SOME DAMAGE

The frit fly, *Oscinella frit* (L.); gout fly, *Chlorops pumilionis* Bjerk.; wheat bulb fly, *Hylemia coarctata* Fall.; lick beetles, Elateridae; carabid beetle, *Zabrus tenebrioides* Goeze.; wheat cockchafer, *Anisoplia segetum* Herbst.; field mouse, *Citellus oken;* and Rustic Shoulder Knot, *Hadena basilinea* F. can all cause some damage to rye crops.

Frit Fly—Oscinella frit L. The frit fly is found throughout Europe and, less frequently, in northern Europe, in northern Asia, and in North America. Damage caused by this pest is substantial, especially in the years when it appears in considerable numbers. It particularly attacks winter rye planted early in the autumn. Besides various forms of *O. frit, O. maura* Fall., *O. anthracina* Meig., *Camarota curvinervis* Latr., and *Elachiptera cornuta* Fall. occasionally are found on rye. The larvae of these other species feed on rye stalk previously infected by *O. frit.*

Affected plants exhibit a characteristic yellowing of the youngest leaf. The stalks do not grow normally and the yellow leaf can be easily removed. The larvae develop at the base of the stem. If the larvae start to feed after heading, the spikes gradually dry up. The fly can also destroy the spikelets in the spike. The larvae do not attack rye grain.

Young frit fly larvae are tapered in front, have bluntly rounded ends, and are 3–4 mm long. The oral hooks which shine through the skin are initially light brown and later black. Mature larva is spindle-shaped, lightly segmented, and black in color. The adult insect is black, 1.5–2 mm long, with a completely black head. Its wings have a metallic sheen and 5 longitudinal and 3 lateral veins.

Adult insects appear in spring and lay eggs on the upper side of leaf blades of cereal plants at the 2–3 leaf stage. After 3–4 days, young larvae hatch and migrate from the leaves to the interior of the plant. The larvae pupate after 15–30 days and after 7–14 days the adult insect appears. The third generation of insects appears in August. Under favorable conditions the frit fly produces 3–4 generations per year.

Some control can be obtained by delaying autumn sowing or by sowing spring crops as early as possible. Wild grasses growing near the rye fields provide a good breeding environment for the insects; these should be eliminated by border cultivation.

WHEAT BULB FLY—*Hylemyia coarctata* Fall.

Hylemyia coarctata Fall. causes similar damage as the green-eyed fly (*Chlorops pumilionis* Bjerk.) except that it eats the whole inner part of the stem and then passes on to a new plant. The damaged plants die and remain standing in the field. Plants which are weakly infected produce too many tillers.

The adult insect is 6–7 mm long and is similar to the domestic fly, yellow-grey with numerous black hairs. The larva is cylindrical, white, and up to 9 mm long. The pupa is yellow-brown and barrel-shaped.

Eggs of the wheat bulb fly winter in the soil. Larvae emerge in early spring and feed on plants. Pupation occurs in the soil. Flies emerge in June and July and after a few weeks they lay eggs.

Rye plants heavily infected by the wheat bulb fly should be plowed deeply to destroy the flies that emerge from pupae.

GOUT FLY—*Chlorops pumilionis* Bjerk.

The Gout fly, *Chlorops pumilionis* Bjerk., occurs in many regions where cereals are cultivated. It attacks rye, especially the late-maturing varieties.

At the heading stage, plants of cereals, especially of wheat and barley, show damaged stems, shortened sub-aerial parts, and spikes that do not grow out of the leaf sheath. On the sub-aerial part, a furrow may be seen. Above the node there is frequently a larva, 6–8 mm long, white, light yellow, or greenish, tapering in front and at the end. Adult insects are 4 mm long and yellow in color. On their heads they have hairs and a black frontal triangle. The abdomen is composed of 5 segments. The Gout fly usually produces two generations per year. The first generation appears during the period from the end of April until June from eggs laid on the upper surface of the leaves. Larvae of this generation cause the greatest losses in rye and wheat. Insects of the second generation lay eggs on plants of winter wheat. They grow into the interior of the plant and locate themselves in the root neck where they winter. The damage in rye crops by this insect may be considerable.

Some control of Gout fly can be achieved by using appropriate periods of planting. In regions where this insect causes great losses, varieties which mature earlier should be used.

CEREAL CHAFER—*Anisoplia segetum* Herbst

The cereal chafer causes losses in rye only in the case of mass appearance. In affected crops, spikes with damaged grain may be found in the field at the stage of milk ripeness. This damage is caused by *A. segetum* beetles which bite into the developing kernels. The cycle of development of the pest lasts two years. Larvae feed on roots of various plants and winter in the soil. The species *A. agricola* Poda, *A. austriaca* Herbst, and *A. lata* Er. can occur with *A. segetum.*

Postharvest plowing to destroy the plant remains and autumn plowing are recommended as the best means of controlling cereal chafer. If the infestation is severe, chemical pesticides should be applied.

RUSTIC SHOULDER KNOT—*Hadena basilinea* F.

Rustic shoulder knot occasionally causes serious losses in rye fields by damaging the grain at the stage of mild ripeness. Adult butterflies appear in May and June, fly by night, and lay their eggs on the leaves of cereals. The caterpillars, which hatch, eat into the soft kernels. During harvest, a considerable number of caterpillars descend on a thread to the ground; some remain in the spikes and are stored with the grain. During storage, the caterpillars can cause considerable grain losses. The caterpillars winter in the soil or under stones. This pest decreases the crop yield and its commercial value.

One method of control is to thresh immediately after harvesting to destroy feeding caterpillars. Immediate plowing and deep autumn plowing are helpful. Caterpillars which accumulate in the area of threshing and/or in the granary should be destroyed.

C. Pests Which May Cause Damage to Rye

Pests which under favorable conditions may multiply considerably and produce some damage to rye crops are: wheat nematode, *Anguina tritici* Steinbuch; cereal bug, *Eurygaster maura* L.; dart moth, *Agrotis segetum* Schiff.; heart-and-dart moth, *Agrotis exclamationis* L.; Hessian fly, *Mayetiola destructor* (Say); leather jackets, *Tipula* spp.; grey field slug, *Agriolimax agrostis* L.; wheat stem sawfly, *Cephus* L.; and rodents, Rodentia.

DART MOTH—*Agrotis segetum* Schiff.

In the climate of central Europe, this pest produces one or two generations per year. It is a multivorous pest. The damage is caused by caterpillars which eat germinating seed and gnaw into leaf blades. The damaged plants break and fall. One caterpillar may destroy several plants. In the zone of feeding, empty areas form in the field. Caterpillars may also destroy young leaves.

Some control may be achieved by proper agronomic practices such as correct first plowing and late autumn plowing. Late planting in the autumn and early planting in the spring should be used. When infection is severe, chemical pesticides should be applied.

CEREAL BUG—*Eurygaster maura* L.

The cereal bug occurs occasionally on rye and causes small economic losses. Adult insects and larvae suck the sap from leaves and stems, causing yellowing and drying up. Plants often form additional side shoots. In later growing periods, the pests draw the sap from spikes and grain. The spikes become white, and the grain is not formed. Flour from partly affected grain cannot be used for baking because of the high proteotylic activity of the injected saliva. *E. maura* produces one generation per year. One female lays about 200 eggs. Larvae feed on leaves and blades. Adult insects appear in May and June and continue to feed on cereals. Before the harvest, they pass on to maize and grasses or fly to wintering places.

For control of cereal bug, early spring planting is recommended as well as the application of fertilizers which speed up the development of plants. Early first plowing after harvesting is sometimes helpful. Chemical agents should be applied in case of mass appearance of the pest.

HESSIAN FLY—*Mayetiola destructor* Say

Under favorable climatic conditions, the Hessian fly can be very dangerous—the more so since caterpillars can maintain the ability for development for up to two years and may be carried with straw from place to place. The degree of infestation depends on the stage at which the plant is attacked. In the case of infestation at an early stage, the plants die or form secondary tillers with weak spikes. Attack of the main tiller is extremely serious since the plant does not form a spike. Affected plants are deformed with abnormally swollen stems and stunted first leaves. The color of the plants is at first intense green and then yellow. Eventually the plants die. The spikes of affected plants are weak and whitish in color. The stems break easily at the lower nodes. At the break, larvae may be found. *M. destructor* winters in the larval stage of the pupa. The adult insect emerges at the end of April or in May. Eggs are laid on the lower leaves. The larvae are deposited at the junction of the stem and the leaf sheath and suck the sap from the stem. In central Europe, the second generation appears in late summer.

Early planting of spring cereals is recommended as one means of controlling the development and spread of Hessian fly. Weeds and self-sown plants should be destroyed. Deep autumn plowing and intense phosphorus-potassium fertilization should be used. In years of mass appearance of the pest, the autumn planting should be delayed.

The above-mentioned pests should be carefully scrutinized by the farmer and appropriate control measures taken to prevent multiplication. The occurrence of some of the pests in rye fields depends on the region and ecological conditions. In southern regions of the rye growing area of Europe, the following pests have been observed: carabid beetles, *Zabrus tenebroides* Goeze; anisoplia beetle, *Anisoplia austriaca* Herbst; pentatomid eurygaster, *Eurygaster integriceps* Put; Souslik, *Citellus oken;* Hessian fly, *Mayetiola destructor* Say; and orthoptera, *Orthoptera.* In the northern regions, the most common are: leather jacket, *Tipula* L.; and green-eyed fly, *Chlorops pumilionis* Bjerk.

PART TWO—PHYSIOLOGY

I. SOME ELEMENTS OF RYE PHOTOSYNTHESIS

The production of rye plant biomass involves photosynthesis, when basic organic molecules and chemical energy are produced, and growth, when these organic building blocks are reorganized to the structures of the plant organism. According to Larcher (1969), photosynthetic productivity of particular plants and ecosystems depends on physiological, morphological, and genetic factors. The biochemical and physiological activity of the plant and its morphological characters are subject to genetic regulation. Environmental factors determine the favorable or adverse conditions for photosynthetic production and in essence determine the distribution of the plants (LAI index) in the ecosystem. The above-mentioned factors, acting on the photosynthetic activity of plants in a system, also influence other physiological processes affecting biomass production, such as the translocation and storage of assimilates and losses due to dissimilation.

Many workers have indicated the possibility of selecting improved plants on the basis of genetic variability of fundamental physiological processes of the plant (Ruebenbauer, 1964). This possibility takes into consideration the genetic differentiation of photosynthetic activity within the plant species and genotypes and the correlation of this activity with crop yields (Apel and Lechman, 1969; Świeżyński, 1972).

Gaastra (1959) and Stoy (1973) showed that the productivity of a field is the result of interactions between physiological and climatic factors in a closed system. Taking into consideration the physiological differentiation of plant components, one should bear in mind the intensity of photosynthesis of various organs, the distribution of the produced assimilates, dissimilatory processes, the size of the assimilatory surface, and the duration of the period of assimilation.

Of the cereal grains, rye is a species to which physiologists have paid relatively little attention. Meanwhile this plant, as an important crop, differs considerably from other cereals, *e.g.*, wheat, in many morphological and physiological characteristics which affect photosynthetic activity and productivity, especially after heading (Birecka *et al.*, 1968b). The stem of winter rye increases in mass for an extended period after heading. During maturation of the grain, the mass of stems diminishes but is distinctly higher than during heading and flowering. The development of the grain is similar to that of winter wheat. It is quite slow; one month after heading, its mass is 1/7 of the final value. This phenomenon has been confirmed by experiments using compounds labelled with ^{14}C. A substantial proportion of the ^{14}C is retained in the stem, as in the case of winter wheat. Participation of the spike in shoot photosynthesis is relatively minor. When the spike of an intact plant was exposed to ^{14}C, about 8% of the total carbon of the seed was labelled. This value did not depend on the age and physiological state of the plant and was similar to that found in winter wheat. The glumes of winter rye have a lower photosynthetic activity than those of other cereal crops grown under similar conditions. The total loss (by respiration or translocation) of ^{14}C from the site of photosynthesis at various stages of growth was 30–36% and was lower only in the case of assimilates formed two weeks before maturation. The incorporation of ^{14}C into assimilates formed later (between the 18th and 35th day after heading) was still high in the case of stems.

Birecka and Zinkiewicz (1968), in investigations on winter rye variety Ludowe, confirmed earlier reports on the Puławskie variety of the low (6–7%) participation of the spike in shoot photosynthesis. It appears that the main factor limiting the relative photosynthetic activity of winter rye spike is the diffusion of CO_2 in the glumes, which in general have fewer stomata per surface area than leaves. In wheat varieties, net CO_2 fixation per mg chlorophyll, soon after heading, was the same in the spikes and sometimes one-half of leaf blades, while in the case of rye it was only one-quarter that of leaves.

Relative to the participation of other green parts of the shoot in photosynthesis of rye, the following points are noteworthy: 1) the minor role of the flag leaf which is of limited surface area relative to lower leaves on the plant; and 2) the important role played by the stem (not only the upper internode but also lower ones), especially after pollination.

In the period after heading, until anthesis, leaf assimilates were transported mainly into the stem and lower parts of the plant. Assimilates formed within 8–9 days after heading were quickly (75% in 24 hr) transported into the stem. This transport continued unchanged until maturity. The stem contained as much as 85% of the total amount of carbon fixed in the shoot. In later stages, a considerable portion of the assimilates was withdrawn into the grain. However, over 45% of the assimilates remaining in the shoot was in the stem.

The participation of the spike, especially before pollination, as the acceptor of assimilates is considerable. Birecka and Zinkiewicz (1968) observed that: 1) removal of the spike had no effect on the flow of assimilates from leaves; 2) when the role of the stem as the acceptor of assimilates was lower, determined by the growth and storing capacity, a higher amount of assimilates was translocated to the spike and roots; and 3) a higher photosynthetic activity of the shoot, in vegetative growth stages as a result of good growing conditions, exerted a strong effect on the assimilates of the spike and sheath in the later stages. This was not related to a higher intensity of storage of assimilates in the shoot but to a considerable decrease in respiration.

The change in ^{14}C distribution observed in rye may be due to the fact that before pollination, when the photosynthetic activity of the shoot is very high, the growth of the stem is not sufficient for it to be an assimilate acceptor (the transporting elements may already be saturated). These data indicate that, in contrast to other grains, the assimilates formed not only before heading but also between heading and pollination do not play an important role in the formation of the biomass of the rye grain.

In both winter and spring wheat, the participation of the flag leaf and the spike in photosynthesis of the shoot is higher than in rye (Strebeyko *et al.*, 1973; Apel *et al.*, 1973). It was shown by Birecka and Dakic-Wlodkowska (1964), Birecka *et al.* (1964; 1968a,b), and Nalborczyk and Nalborczyk (1973) that translocation of assimilates from leaves into the spike and the utilization of assimilates in the sheath and spike are important in grain formation. The participation of spike assimilates in grain development is 15–29% and depends on rye variety and stage development.

It is interesting to compare rye with other cereal plants. The participation of the flag leaf in the yield of barley is considerable (Berdahl *et al.*, 1972; Thorne, 1963a) and, together with the photosynthesis of the sheath and leaf blade, amounts to about 50–60% of the total. Photosynthesis of the spike and the participation of spike assimilates are higher in barley than in wheat or rye. Apel (1967), Birecka *et al.* (1964), Thorne (1963b), and Watson *et al.* (1958) indicated that the spike plays a considerable role in barley grain production. The participation of the products of spike photosynthesis in the amount of assimilates accumulated in the grain was 33–35%. This value agrees with the data of Thorne (1963a), who estimated the contribution of assimilates formed by the spike to be about 40% of the final grain yield. These values are considerably higher than those found for rye (15–29%).

II. REACTION TO DROUGHT

Rye is an undemanding plant; it can be grown on dry soils of moderate fertility. Of all the cereals, rye has the most highly developed root system. The roots take up water and nutrients very efficiently. Rye uses 20–30% less water per unit dry matter formation than wheat. Investigations on the water balance in rye have not been as numerous as on wheat. Listowski and Domańska (1960) noted that a small water deficit in autumn decreased the susceptibility of rye to drought in spring.

Distinct differences in resistance to drought exist between rye varieties. Strebeyko and Domańska (1954) proposed that the water-absorbing capacity of germinating rye seeds be examined in order to determine the degree of sensitivity and resistance to deficits of water.

Slaboński and Lipińska (1969) found that water deficits in the period from shoot formation until flowering and from heading to milk maturity are the most harmful to grain yield. Drought in the period from tillering to heading had relatively little effect on yield. Tetraploid rye varieties are more sensitive to drought than diploids. On dry soils with a continuous water deficit (about 40% of field capacity) rye may give higher yields than on soils where water is abundant but periods of drought occur. Differences also exist between varieties in their reactions to particular periods of drought.

Hallam *et al.* (1972) distinguished three stages in the water uptake by rye seeds. The first lasts 10 min; the second is slower and lasts about 1 hr. During the second stage, changes in cell ultrastructure can be observed, especially in number of mitochondria. The third stage, active water uptake, is characterized by a proliferation of endoplasmic reticulum which gradually becomes more dense around the cell nucleus after 6 hr. Protein synthesis begins during the third stage of water uptake.

III. NUTRITION OF THE RYE PLANT

Winter varieties of rye are grown in most of the rye-cultivating regions. The seed is planted in the autumn; it germinates and grows into a plant of about 15 cm. The young plant remains essentially dormant during the winter, usually covered by snow, and then continues to grow to maturity in the spring and early summer. The autumn weather is characterized by low temperatures of both soil and air and by significant daily changes in temperature. Under such conditions mineral nutrients, especially phosphorus and nitrogen, are extremely important. The role of the more important mineral nutrients in rye growth will be discussed.

A. Phosphorus

Plants growing under conditions of limited available phosphorus in soil do not absorb sufficient nitrogen and show distinctive symptoms of nitrogen deficiency. The rate of nitrogen absorption and the degree to which plant requirements for this element are satisfied depend on the type of nitrogen fertilizer. Plants absorb nitrogen either as nitrate or as ammonium. Nitrate is absorbed more slowly than ammonium. Plants absorbing nitrogen in the form of nitrate show a greater demand for phosphorus. This phenomenon is particularly evident in the autumn when the temperatures of soil and air are lower. The effects of low temperatures can be indirectly limited if the quantity of phosphorus that plants absorb in this period is increased. Phosphorus increases the level of available energy for uptake and assimilation, thus facilitating a more intensive uptake of nitrogen and its incorporation into nitrogen compounds.

The interdependence between uptake of phosphorus and of nitrogen in autumn has been observed in rye. Bezludnyi and Belenkevich (1970), using the winter variety of rye Benyakowskaya grown in pots, showed that the nitrogen uptake by the plant depended on the amount of phosphorus in the soil. A similar dependence, though not to the same extent, was observed for roots. In addition, lack of phosphorus decreases the translocation of nitrogen from the roots to the aboveground parts of the plant (Table II). Under conditions of sufficient phosphorus in the soil, the uptake of nitrate is similar to that of ammonium. Shortage of phosphorus appears to limit the absorption of nitrogen in the form of nitrate. Under these conditions, the synthesis of nitrogen and protein compounds is slower. The data of Bezludnyi and Belenkevich (1970) illustrate the dependence of nitrogen absorption from various fertilizers in the presence and absence of adequate phosphorus, as well as the distribution of nitrogen among various compounds (Table III).

Table III shows that, in the case of phosphorus deficiency, the intake of nitrate nitrogen is lower by a factor of two than that of ammonium nitrogen. Similar results were obtained from the analyses of nonprotein compounds as from the analyses of free and bound proteins. Accumulation of the dry mass of rye is closely related to nitrogen absorption (see Table IV).

Data of Table IV also show that absorption of ammonium nitrogen is higher than nitrate nitrogen. However, the differences here are smaller than in Table III. According to the authors, this results from limited sugar synthesis due to insufficient artificial light used to simulate autumn conditions.

B. Potassium

A deficiency of potassium in the soil results in a limited absorption of other nutrients. Baier and Smetánková (1968) examined the relation of the contents of nitrogen, phosphorus and potassium, ratios $N:K_2O$ and $N:P_2O_5$ in straw and in grain, and the relation between grain yields and mineral nutrient uptake from the soil in the periods to flowering and to maturity. A deficiency of potassium distinctly decreased the uptake of other nutrients by the rye plant. This effect was particularly evident during the periods of stem and leaf development. Nitrogen plays the major role in the grain-forming process.

The negative effect of potassium deficiency on the utilization of absorbed minerals and formation of grain is greater in side shoots (younger plants) than in the main shoot. Smetánková and Baier (1968) suggested that in rye the nutrient components are mobilized from the youngest shoots to the main shoot. Both the quantity and nature of nutrient components play a considerable role in grain formation and the synthesis of storage proteins of the grain endosperm. A complete NPK fertilizer contributes to the increase of both the weight of grain (yield) and the grain protein content. Nutrients that increase the acidity of the soil disturb the synthesis of proteins in the rye plant. Application of N and P fertilizers increases the accumulation of new protein nitrogen substances. This process may be diminished by the addition of lime (soil liming). This is rather surprising since rye, of all the cereals, has the greatest tolerance to variations of soil pH. Determinations of the contents of various proteins in rye flour showed that gliadin content depends significantly on the mineral nutrient level and balance (Titova, 1969). The contents of albumin, globulin and glutenin did not show significant fluctuations.

TABLE II

Effect of P on N incorporation into various plant parts

Nitrogen Source	P_2O_5 mg/pot	Above-Ground Parts			Roots	
		Non-protein N	Structural protein N	Soluble protein N	Non-protein N	Protein N
		mg				
$Na^{15}NO_3$	21.3	20.4	27.9	21.6	10.2	14.8
$Na^{15}NO_3$	213.0	41.3	54.1	26.4	11.4	20.0
$(^{15}NH_4)_2SO_4$	21.3	27.3	35.9	24.1	17.1	25.8
$(^{15}NH_4)_2SO_4$	213.0	45.8	54.7	30.4	19.1	25.1
		Degree of Isotope Enrichment (^{15}N)				
$Na^{15}NO_3$	21.3	2.11	1.04	0.97	2.41	1.23
$Na^{15}NO_3$	213.0	4.32	2.97	2.90	4.50	2.77
$(^{15}NH_4)_2SO_4$	21.3	4.42	2.97	2.90	4.50	2.77
$(^{15}NH_4)_2SO_4$	213.0	5.90	3.39	3.88	5.19	3.00

TABLE III

Effect of P on N incorporation into various plant parts

Nitrogen Source	P_2O_5 mg/pot	Above-Ground Parts				Roots		
		Non-protein N	Soluble protein N	Structural protein N	Total N	Non-protein N	Protein N	Total N
		mg						
Nitrate	21.3	3.47	1.27	1.80	6.54	2.07	1.25	3.32
Nitrate	213.0	16.0	9.2	14.00	39.40	5.93	4.78	10.71
Ammonium	21.3	11.0	3.51	6.40	20.91	7.04	4.33	11.37
Ammonium	213.0	25.2	7.60	16.40	49.2	9.18	6.57	15.75

TABLE IV

Effect of P on N incorporation into different plant parts

P_2O_5 mg/pot	$Na^{15}NO_3$		$(^{15}NH_4)_2SO_4$	
	Above-ground parts	Roots	Above-ground parts	Roots
21.3	1.80 ± 0.10	1.35 ± 0.14	1.90 ± 0.04	1.90 ± 0.08
213.0	3.47 ± 0.08	1.52 ± 0.17	2.96 ± 0.11	1.95 ± 0.07

C.   Copper

Pichl and Ružička (1970) showed that the copper content of various cereals is approximately the same. The average copper content has been found to be 0.4 mg/100 g for all cereals, 0.48 mg/100 g for rye, and 0.33 mg/100 g for barley.

A separate and a more complicated problem is the balancing of the quantities of nutrient components absorbed by plants from soil, as compared with yields and contents of major elements in grain. Bogusławski and Gierke (1961) showed that the higher the yield, the larger the quantity of NPK absorbed by plants (grain and straw). Chojnacki and Boguszewski (1971) compiled data (Table V) which indicate that this relation exists in rye. These authors concluded that the NPK content of plants depends on many factors, some that can and some that cannot be controlled by the farmer.

Generally speaking, under the conditions of continental climate, the nitrogen content of plants is higher than under more moist conditions. The changes in weather conditions in some years provoke fairly high oscillations in the content of some mineral components. In some cases the climatic conditions exert a greater influence than the type of soil. An increase in fertilizer addition, especially nitrogen, has a marked effect on yield.

IV.   GROWTH AND DEVELOPMENT

Populations of cultivared rye consist of winter, spring, and intermediate phenotypes. Spring rye planted in the autumn can change into the winter form as a result of natural selection and adaption processes. The period of vernalization of spring rye is relatively short (10–12 days) while that of winter forms is longer (40–60 days). There are many biotypes which are intermediate between these two extremes.

Rye is a long-day plant. Bühring (1960, 1965) found that shortening the day to 9 hr causes a delay in plant development and inhibits shooting. Rye plants subjected to short photo periods resemble grasses and may persist in this state for up to seven years.

The ontogenesis of the rye plant was investigated by Kuperman et al. (1955) and Kuperman (1962). The developmental cycle was divided into 12 stages. During stage I, the growing point is not differentiated. In stage II, primordia of stems, nodes and internodes are formed in the growing point. Winter rye planted in the autumn in a moderate climate enters the winter period in stage II. In stage III, the growing point differentiates into further segments which are primordia of spikelets. In this period, compounds containing available nitrogen in the soil have a positive effect on the formation of a large number of spikelets which leads to the subsequent formation of longer spikes with a greater number of flowers and grains. A further differentiation of growing points takes place during stages III and IV in which flower primordia are formed. This process takes place in early spring.

During the formation of spikelet primordia in the upper part of the spike, organs of generative reproduction (flowers) are formed in the middle portion. The plants then enter stage V of organogenesis. Under conditions of long-day and with a poor nitrogen supply, this process is relatively fast.

Meiotic divisions of pollen mother cells, and the formation of tetrads, the embryo sac, and the egg take place during stage VI and VII of organogenesis.

TABLE V

Changes in the percentage of components in rye caused by fertilizing

Authors	Lower Level of Fertilization									Higher Level of Fertilization								
	Doses in kg/ha			Per cent						Doses in kg/ha			Per cent					
				N		P_2O_5		K_2O					N		P_2O_5		K_2O	
	N	P_2O_5	K_2O	Grain	Straw	Grain	Straw	Grain	Straw	N	P_2O_5	K_2O	Grain	Straw	Grain	Straw	Grain	Straw
Boguszewski and Gosek (1971)	...	36	60	...	...	0.68	...	0.43	...	...	72	120	...	...	0.73	...	0.43	...
Boguszewski, Gosek, Grześkiewicz (1971)	...	36	40	...	...	0.72	0.21	0.43	1.18	...	72	160	...	...	0.74	0.22	0.45	1.28
Boguszewski, Gosek and Zaleski (1971)	...	36	60	...	...	0.77	0.25	0.59	1.46	...	72	120	...	...	0.80	0.28	0.60	1.64
Boguszewski, Maćkowiak and Maćkowiak (1961)	40	27	60	1.16	0.38	0.85	...	0.49	1.50	80	54	120	1.20	0.39	0.80	...	0.52	1.74
Boguszewski and Pentkowski (1969)	30	...	...	1.68	...	...	...	...	...	60	...	...	1.71	...	...	...	...	...
Goralski and Mercik (1967)	45	...	...	1.48	0.39	...	...	...	...	75	...	...	1.57	0.44	...	...	...	...
Klupczyński (1967)	45	...	...	1.24	0.34	...	...	...	...	90	...	...	1.43	0.42	...	...	...	...
Kuszelewski (1965)	30	30	40	1.45	0.52	0.63	0.24	0.67	1.48	70	45	60	1.51	0.50	0.67	0.20	0.71	1.44
IUNG[a] Data	20	18	30	1.50	0.49	0.96	0.28	0.55	1.52	60	54	90	1.60	0.51	1.01	0.29	0.58	1.74

[a]Institute of Soil Science and Plant Cultivation, Puławy, Poland.

Meiotic divisions begin in the middle part of the spike and gradually proceed upwards and downwards in the spike, usually a few days before the emergence of the spike from the leaf sheath.

The frequency of meiotic divisions of pollen mother cells is highest in anthers which are yellow-green in color (Starzycki, unpublished) and are the most suitable for fixing. To ascertain whether there are dividing mother cells in a given spike, one can make a preliminary check for cell divisions under a microscope after treating with a 2% solution of aceto-carmine.

Period VII is characterized by extensive elongation growth during which pronounced elongation of shoot internodes takes place. In the next stage, VIII, the plants ear and subsequently flower. Fertilization and maturation of caryopses and plants then follow in the remaining four stages of development.

Knowledge of the developmental processes permits determination of the developmental rhythem of varieties and cultivated races, the length of particular stages, and the vegetative period of the cultivated material.

Many physiological and biochemical processes take place which affect the plant growth and development between genetic information and its realization in the form of harvest yield (Ozbun and Wallace, 1974). The regulation of the course and order of metabolic changes implicit in the genetic code reaches the molecular level and is apparent in morphogenesis as processes of biosynthesis of mainly enzymatic proteins (Kirk and Jones, 1970; Konarev, 1972).

Hallam (1972) and Hallam *et al.* (1972) reported that ultrastructural and related biochemical changes take place during embryogenesis and early stages of rye germination. In these investigations, cells of young embryos were found to have highly organized organelles. During maturation and dehydration the complex is considerably simplified. Many layers of the endoplasmic reticulum are reduced; lipid droplets migrate to closed plasmalemma cisternae and mitochondrial cristae are reduced. These structural changes are accompanied by limitation of synthesis and respiration.

In embryos of monocotyledonous plants, protein synthesis starts after 30 min of water imbibition, but nucleic acid synthesis is delayed for several hours. Rye embryos absorb water and return to conditions that ensure a normal structure and course of biochemical processes. An increase in respiration intensity and an associated increase in the number of mitochondria and cristae take place. In later stages of water imbibition (6 hr) the endoplasmic reticulum is extensively developed and uptake of labelled uridine and thymine can be detected. This suggests that protein synthesis starts on new ribosomes associated with the reticulum. During the first 6 hr of water absorption, protein synthesis must take place on polysomes. Roberts *et al.* (1972) and Roberts and Osborne (1973) consider that inactivation of labile transferase I may be the most important factor in aging and loss of viability in rye seeds. In poorly viable embryos, although tRNA activity is not affected, no activity of transferase I is noted and the decrease of protein synthesis *in vivo* is related to this lack of activity.

Feierabend (1969, 1970), in investigations of the formation of photosynthetic enzymes in rye seedlings, found that these processes are independent of light induction but are strongly affected by cytokinins. The development of the photosynthetic apparatus is related to characteristic changes in enzymes that are not functional components of photosynthesis. The appearance of photosynthetic enzymes in the first leaves causes a decrease in the activity of enzymes of the pentose-phosphate cycle. This is due to specialized biological regulation. In rye seedlings, an appropriate application of kinetins and auxins affects glucose-6-phosphate dehydrogenase.

Reducing enzymes of the pentose-phosphate cycle, similar to carboxydismutase (EC 4.1.1.39) and phosphoglyceraldehyde dehydrogenase (EC 1.2.1.9), are formed in the first leaves in etiolated rye seedlings. The extent of their synthesis is determined by the cytokinin level. After adding kinetin to seedlings growing in the dark, those enzymes studied attained the same activity as did those of plants growing in light without the addition of kinetin. The formation of photosynthetic enzymes may be strongly reduced by cutting off the roots in early developmental stage and thereby decreasing cytokinin supply. The addition of kinetin restores synthesis to the previous level. Changes in cytokinin content affect the formation of those enzymes investigated. Other enzymes do not show such a strong dependence on the cytokinin level. Cytokinins seem necessary for the formation of enzymes of the pentose cycle; however, they do not affect the time of appearance of these enzymes. The action of cytokinins may determine specific gene activity. Phytochromes also affect the synthesis of photosynthetic enzymes in a manner independent of cytokinins.

After destruction of the roots, carboxydismutase and phosphoglyceraldehyde dehydrogenase exhibit a higher activity under red and blue light than under far red light. Chlorophyll is not formed under the latter light condition. There appears to be a correlation between the synthesis of these enzymes and chlorophyll. In later work, Feierabend (1970) investigated the interaction of kinetins, auxins and specific inhibitors of protein synthesis.

The growth of plastids, like the synthesis of photosynthetic enzymes, is selectively lowered in the dark or in seedlings growing in light after treatment with relatively high concentration of auxins of the 3-indoleacetic acid type, such as α-naphthaleneacetic acid or 2,4-dichlorophenoxyacetic acid. This inhibition can be eliminated by the addition of cytokinins together with the auxin. On the other hand, the enhanced effect of kinetin on the synthesis of photosynthetic enzymes can be reduced by simultaneous application of α-naphthaleneacetic acid. In older seedlings this reverse effect occurs more rapidly.

The later stages of plant development in respect to metabolic changes are no less interesting than the first stages of the vegetative phase. Vernalization and photo-periodism are two phenomena in which external conditions affect plant development by induction of the generative stage of development. Thomson and Zalik (1973) reported that the content of unsaturated fatty acids in plants increases when they are grown in low temperature. Mitochondrial lipids from cold-resistant plants were more highly unsaturated than those from sensitive plants. Results of model experiments performed *in vivo* and *in vitro* suggest that the levels of unsaturated fatty acids and the polarity of lipid groups affect the physical state of membranes and the activities of enzymes associated with these membranes. In the winter rye variety Sangasto the total phospholipid content in the dry mass was higher than in the spring variety Prolific. During periods of cold, the contents of lipids in the two varieties were similar. Other authors (Markowski *et al.*, 1968; Markowski and Dubert, 1972) reported metabolic changes in metabolism of phosphorus compounds, especially in nucleic acids and changes in ribonuclease activity in vernalized and unvernalized forms of winter wheat.

The work of Evans *et al.* (1973) indicates that spring rye varieties have higher amounts of alkylresorcinols. These compounds have been implicated as the cause of lower palatability of rye as compared with other cereals in animal feeding. Schweizer (1968) investigated the effects of simazine on protein content and found that in field experiments simazine, applied to the soil at a rate of 1/32 to 4 lbs/acre, caused an increase in the protein content of rye and other plants. This herbicide, in doses of 0.5 lbs/acre, increased the protein content in rye and fodder by 44 and 28%, respectively. These results have not been confirmed.

V. EFFECT OF ADVERSE ENVIRONMENTAL FACTORS

Compared with other winter cereals, rye is the most cold-resistant. However, in most areas where winter rye is grown there are significant losses due to the combined effects of snow and low temperatures (below $-25°C$). In addition, winter losses of plants can result from pathogenic fungi, excessive moisture, or ice crust formation. Damage to winter rye is sometimes more serious than to wheat because the rye plant is much taller in the autumn.

The wintering organ in rye, like in other cereals, is the node tiller, which is able to regenerate the state of tillering and rooting of the plant.

In autumn, properly developed rye plants have root meristems able to form large numbers of secondary meristems. Rye requires at least 55 days of autumn vegetation with total heat units of 350–550 degree days. A strongly developed leaf mass causes conditions unfavorable to good wintering by delaying the freezing of the soil. The amount of stored substances used for respiration depends on the leaf mass. Under these conditions, hypoxia and carbon dioxide accumulation under the snow/ice layer in early spring has been observed (Levitt, 1956; Procenko and Kolosha, 1969).

An important role has been attributed to the contribution of pigments to the physiology of plant resistance to low temperatures. Even though the role of plastids has not been definitely determined, it is known that their structure changes when temperatures are lowered. The pigment content of the leaves depends on the temperature. The biosynthesis of pigments and photosynthesis in various plants depends on the soil and air temperatures. In plants which have assimilating organs throughout the whole year and which grow in areas where temperatures fall below $0°C$, the biosynthesis of chlorophyll may take place at temperatures as low as $-2°C$, and that of carotenoids as low as $-5°C$.

The amount of pigment in plant leaves decreases from autumn to winter. In hardy plants, this process takes place to a lesser extent than in plants with low cold resistance (Rybakova and Denisova, 1972). The decrease in pigment content depends on the given conditions of wintering and varies from year to year.

In autumn, the pigment (mainly chlorophylls a and b) content of rye and wheat leaves gradually increases to a maximum at the end of autumn. In general, the leaf pigment content of different varieties of rye and wheat is similar. However, the amount of decrease from autumn to winter is different for different genera and varieties. The pigment drop for winter rye is quite small. This is probably related to the functioning of chloroplasts in this period. The high content of pigment in rye leaves in early spring is implicated in the resistance of the rye plant to low winter temperatures.

VI. BIOLOGY OF SEEDS

A. Germination

The phenomenon of post-harvest germination of cereals was described more than 70 years ago. Numerous scientists have worked on this subject. The length of the dormancy period of rye grain depends on climatic factors. In rye grown under conditions of abundant rainfall and relatively high air temperatures during the period of formation and maturation of grain, the dormancy period of the caryopses is not stabilized and the seeds may even germinate in the ear. In years when harvest temperatures are high and rainfall is sparse, the seeds have a low tendency to germinate during harvest. Several days after harvest the germinating ability increases sharply. In other cereals (*e.g.*, barley), the climatic conditions during the dormancy of seeds have only a slight effect on the length of the dormancy period. High temperatures and humidity may even lengthen the dormancy period of the seeds.

According to Kummerov (1965), the high germinating ability attained by rye seeds does not mean that this ability will remain at a constant level. There exists a possibility of cyclic variation in the germinating capacity of seeds. This hypothesis was checked by Mazurek and Mazurek (1971). The rye variety used has a short dormancy period. The length of the period was decreased if, during maturation, high air temperatures and abundant rainfall occurred. Under conditions of high rainfall and low temperatures, the dormancy period was longer. If the rainfall was too sparse from the time of milk maturity until full maturity, the period of dormancy was shorter. The germinating ability of rye seeds varied in a cyclical manner in the period from harvest until planting. On the basis of these results, the authors have recommended that if low germination is obtained, the germination test should be repeated every few days to ensure that every part of the cycle is examined.

In the soil, rye germinates in a manner similar to other cereals. The lowest temperature at which rye will germinate is $3°-5°C$. The optimum is $25°-31°C$, slightly higher than for wheat.

B. Vernalization

The term vernalization is applied to the treatment of young plants by the low temperature to induce a shorter vegetative period and hasten flowering and fruiting. In cereals, treatment temperatures range from $-2°$ to $+10°C$. The optimum vernalization varies slightly among cereals. The number of days of low-temperature treatment that is necessary to induce the required developmental changes also varies among cereals. There is an interaction between the number of days of vernalization and day length. If vernalization takes place at low temperatures, the day length has little effect. On the other hand, at higher temperatures, day length distinctly affects the results. Interesting observations in this regard on rye were made by Listowski (1955). The period of vernalization in winter rye terminates before winter or at its beginning (Table VI). Vernalization treatment can be reversed by high temperatures (Purvis and Gregory, 1952).

TABLE VI
Effect of length of vernalization period on heading

Variety	Vernalization (days)	Day Length[a]	Spike Formation Days after Vernalization	Maximum Heading Days after vernalization	Maximum Heading Spikes	Spikes Per Plant (average)	Headed Plants %	Straw g dry
Petkus	42	L	39	80	110	1.54	100	25.8
	42	K	57	144	48	0.67	62.5	22.1
	21	L	52	205	60	0.83	80.6	30.0
	21	K	80	205	32	0.44	44.5	24.2
	0	L	125	205	11	0.15	15.3	32.4
	0	K	125[b]	125	2	0.03	3.0	22.6
Vyatka	42	L	37	80	101	1.40	100	20.6
	42	K	55	144	44	0.60	60	17.9
	21	L	46	144	71	0.98	81.9	23.3
	21	K	80	205	37	0.50	51.4	27.3
	0	L	only vegetative development (205 days)				0	35.2
	0	K	125	125[b]	1	...	1.3	23.7

[a] L = long-day; K = short-day.
[b] Only vegetative development except for one plant.

C. Autumn Growth and Development

The first, and to some extent the second, leaf of rye has the ability to accept and transfer to the higher node the light stimuli of differentiation of the tiller and the meristems of secondary shoots (Langer, 1963). Growth of the young plant takes place almost exclusively from material stored in the seed. The second and third leaves provide assimilates for the tiller, but the amount of assimilates provided is insufficient for growth of lateral buds. Lowering light intensity and shortening the day to less than 12 hr affect the tillering processes adversely. The loss of any of the first three leaves unfavorably affects tiller formation and decreases the number and viability of lateral meristems. The amount of assimilates is also lowered and this also delays tillering. The fourth leaf normally develops in late autumn. At this time, climatic conditions inhibit growth more than they do assimilation. Thus, the loss of the fourth leaf has no great effect on growth. Because of this, it is possible to graze or mow rye plants in late autumn without decreasing the grain yield (Jackowska, 1971).

D. Fertility

Rye is a plant that is rather sensitive to adverse weather conditions. If, during the period of flowering, the weather is cold and there is much rain, seed set is not satisfactory. Under such conditions, the spikes have many empty florets. This phenomenon is caused both by adverse conditions of the habitat and by genetic factors.

Another reason for poor seed set may be a decreased viability of the pollen. The sterility of pollen is slightly higher in tetraploid than in diploid forms. According to Tarkowski (1966), both diploid and tetraploid rye varieties differ considerably in this respect. A distinct lowering of the fertility of plants causes sparse seed set in lateral shoots which are poorly developed. If the fertility of main shoots is taken as 100%, it is about 10% lower in shoots 5 and 6, 20% lower in shoot 10, and in additional shoots (15 and 16) seed set can be as low as 50%.

An analysis of seed set at particular levels in the spike has also been carried out. The set is best in the middle of the spike and worst at the first and second levels at the base of the spike and at the topmost levels.

Tetraploid rye in general has a smaller number of spikelets per spike than diploid rye. Some of the flowers of the lower levels of spikes in tetraploid rye are abscised just after flowering. Weaker setting of seeds in tetraploid rye is due to a tendency to form aneuploids, a smaller number of flowers in spikelets, and perturbations in the development of the embryo and endosperm. These phenomena occur to a greater degree in varieties than in hybrids (Müntzing, 1954, 1963; Słaboński, 1964).

Ruebenbauer and Müller (1963) consider that weather factors in the period of flowering of rye are of utmost importance in increasing fertility. Moore (1963) found that there is no simple correlation between the course of meiotic divisions and of seed setting. He considers that monovalents are the main cause of perturbations in meiosis even though some physiological factors may increase aneuploid formation.

LITERATURE CITED

ANONYMOUS. 1953. Some further definitions of terms used in plant pathology. Trans. Brit. Mycol. Soc. 36: 267.

APEL, A., TSCHÄPE, M., SCHALDACH, I., and AURICH, O. 1973. Die Bedeutung der Karyopsen für die Photosynthese und Trockensubstanzproduction bei Weizen. Photosynthetica 7: 132–139.

APEL, P. 1967. Potentielle Photosynthese-intensität von Gerstensorten des Gaterslebener Sortiments. Kulturpflanze 15: 171–174.

APEL, P., and LECHMAN, C. O. 1969. Variabilität und Sortenpezilität der Photsyntheserate bei Sommergerste. Photosynthetica 3: 255–262.

ATANASOFF, D. 1923. Fungi blight of the cereal crops. Meded. Landb. Dell. 27: 1–132.

BAIER, J., and SMETÁNKOVÁ, M. 1968. Concentration and nutrient ratio in winter rye with ears of differing size. Rostl. Výroba. 14: 773–792.

BALTZER, V. 1930. Untersuchungen über Anfälligkeit des Roggens für Fusariosen. Phytopath. Z. 2: 337–341.

BENADA, J. 1963. Some properties of the orange red pigment of snow mould Fusarium nivale (Fr.) Ces. and its diagnostic value. Sc. Mycologie 17: 98–101.

BERDAHL, J. D., RASMUSSON, D. C., and MOSS, D. N. 1972. Effect of leaf area on photosynthetic rate, light penetration and grain yield in barley. Crop Sci. 12: 177–180.

BEZLUDNYI, N. N., and BELENKEVICH, O. A. 1970. The nitrogen metabolism in winter rye plants as affected by conditions of phosphorus nutrition and nitrogen forms. Fizj. Rast. 17: 992–996.

BIRECKA, H., and DAKIC-WLODKOWSKA, L. 1964. Photosynthesis, translocation and accumulation of assimilate in cereals during grain development. Acta Soc. Bot. Pol. 33: 407–426.

BIRECKA, H., SKUPINSKA, J., and BERNSTEIN, J. 1964. Photosynthesis, translocation and accumulation of assimilate in cereals during grain development. Acta Soc. Bot. Pol. 33: 601–618.

BIRECKA, H., WOJCIESKA, H., and GLAZEWSKI, S. 1968a. Ear contribution to photosynthetic activity in winter cereals.

I. Winter wheat. Bull. Acad. Pol. Ser. Sci. Biol. 16: 191–196.

BIRECKA, H., WOJCIESKA, H., and ZINKIEWICZ, E. 1968b. Ear contribution to photosynthetic activity in winter cereals. Part II. Winter rye. Bull. Acad. Pol. Sci. Ser. Sci. Biol. 16: 257–260.

BIRECKA, H., and ZINKIEWICZ, E. 1968. Photosynthesis and ^{14}C-assimilate translocation in winter rye after heading. Bull. Acad. Pol. Sci. Ser. Sci. Biol. 16: 323–329.

BOGUSŁAWSKI, E., and GIERKE, K. 1961. Neue Untersuchungen über den Nähr-Soffentzug verschiedener Kulturpflanzen. Z. Acker Pflanzenbau 112: 226–252.

BOGUSZEWSKI, W., and GOSEK, S. 1971. Effectiveness of phosphorus and potassium fertilizing on the ground of regional experiments. Pamiet. Puławski 50: 29–41.

BOGUSZEWSKI, W., GOSEK, S., and GRZEŚKIEWICZ, H. 1971. The results of experiments with high doses of phosphorus and potassium in experimental stations of Institute of Soil Science and Cultivation of Plants. Pamiet. Puławski 42: 55–78.

BOGUSZEWSKI, W., GOSEK, S., and ZALESKI, T. 1971. Results of experiments with high doses of phosphorus and potassium in experimental stations of Institute of Soil Science and Cultivation of Plants. Part II. Pamiet. Puławski 50: 43–52.

BOGUSZEWSKI, W., MAĆKOWIAK, Cz., and MAĆKOWIAK, W. 1961. Effects of mineral fertilization ration in a four course plant on sandy soil. Pamiet. Puławski 2: 221–234.

BOGUSZEWSKI, W., and PENTKOWSKI, A. 1969. Investigation concerning the term of fertilizing rye with additional nitrogen rates. Pamiet. Puławski 37: 137–148.

BOJARCZUK, J. 1968. Testing rye varieties for stem break (Cercosporella herpotrichoides Fron.) Hodowla Rośl. Aklim. Nasiennictwo 12: 645–656.

BOJARCZUK, J., and BOJARCZUK, M. 1972. Evaluation of susceptibility to stem break (Cercosporella herpotrichoides Fron.) in new rye varieties and strains. Biul. IHAR, 3-4: 67–71.

BRUNDZA, K. 1943. Der Schneeschiemmel in Litauen. Angew. Bot. 25: 112–113.

BÜHRING, J. 1960. Beobachtungen und Untersuchungen am Winterroggen über Schossverzögerung und Erhaltung von

Klonen durch photoperiodische Behandlung. Z. Pflanz. Zücht. 43: 266–296.

BÜHRING, J. 1965. Uber die photoperiodische Kurztagbehandlung von Winterroggen. Z. Pflanz. 48: 134–135.

CHOJNACKI, A., and BOGUSZEWSKI, W. 1971. Content of nitrogen, phosphorus and potassium fertilizing on the ground of regional experiments. Pamiet. Puławski 50: 5–27.

CZARNOCKA, J. 1939. The rye variety Puławskie Wczesne and methods of its breeding. PINGW, Puławy.

ECOCHARD, R. 1963. Charactéristiques genétiques de certains Triticum d'Ethiopie résistans à *C. herpotrichoides* et à *P. graminis*. Ann. Amelior. Plant. 13: 5–25.

EVANS, L. E., DEDIO, W., and HILL, R. D. 1973. Variability in the alkylresorcinol content of rye grain. Can. J. Plant Sci. 53: 485–488.

FEIERABEND, J. 1969. Influence of cytokinins on the formation of photosynthetic enzymes in rye seedlings. Planta 84: 11–29.

FEIERABEND, J. 1970. Characterization of cytokinin action on enzyme formation during the development of the photosynthetic apparatus in rye seedlings. Planta 94: 1–15.

GAASTRA, P. 1959. Photosynthesis of crop plants as influenced by light, carbon dioxide, temperature and stomatae diffusion resistance. Meded. van de Landbouwhogesch. te Wageningen Nederland 59: 1–68.

GARBINI, S. E. 1950. Comportamiento de vanedades de avena cebada y centeno ensayades en Pergamino. Per. exp. St. Ann. Rep. Agr. Exp. St. Univ. Florida. 19–21.

GARBOWSKI, L. 1928. Plant diseases, Ksieg. Rolnicz. Warszawa.

GÄUMANN, E. 1959. Die Rostpielze Mitteleuropas. Beitr. Kryptogamenfl. Schweiz. 12: 117–118.

GERM, H. 1960. Das Problem der Keimfahigkeitsbestimmung von fusarium — kranken Weizen — und Roggenssatgut. Saatgutwirt. 12: 164–165.

GORALSKI, J., and MERCIK, S. 1967. Effect on growing rates of urea top dressing for cereals. Rocz. Nauk Roln. Ser. A 93: 263–285.

HALLAM, N. D. 1972. Embryogenesis and germination in rye (*Secale cereale* L.). I. Fine structure of the developing embryo. Planta 104: 157–166.

HALLAM, N. D., ROBERTS, B. E., and OSBORNE, D. J. 1972. Embryogenesis and germination in rye (*Secale cereale* L.). II. Biochemical and fine structure changes during germination. Planta 105: 293–309.

JACKOWSKA, J. 1971. The role of autumn development of plants from early sowing for overwintering and productivity of winter rye. Pamiet. Puławski 47: 55–77.

JAMALAINEN, E. A. 1959. Overwintering of Gramineae plants and parasitic fungi. J. Sci. Agr. Soc. Finl. 31: 282–284.

KIRK, D., and JONES, R. N. 1970. Nuclear genetic activity in B-chromosome rye in terms of the quantitative interrelationships between nuclear protein, nuclear RNA and histone. Chromosoma 31: 241–254.

KLINKOWSKI, M., MUHLE, E., and REINMUTH, E. 1965. Phytopathologie und Pflanzenschutz. Akademie Verlag, Berlin.

KLUPCZYŃSKI, Z. 1967. Influence of nitrogen fertilizing on the yield of rye and winter wheat and on the content and composition of protein in the grain. Pamiet. Puławski 24: 229–250.

KOBYLYANSKII, V. D., and ILICHEV, G. A. 1972. The initial plant material for diploid winter rye breeding. Sel. i Sem. 37: 35–39.

KOCHMAN, I. 1973. Phytopathology. PWRiL, Warszawa.

KONAREV, V. 1972. Molecular biology and some problems of plant growing. Vest. Sel'skokhoz. Nauki 17: 7–13.

KUMELOWSKA, I. 1961. Some problems of ergot breeding, growing and processing. Biul. IHAR 6: 93–96.

KUMMEROV, J. 1965. Enc. of Pl. Phys. XV, 721–726.

KUPERMAN, F. M. 1962. The biological check-up in agriculture. Moscow University, Moscow.

KUPERMAN, F. M., DWORIANKIN, F. A., RŻANOWA, E. Y., and ROSTOWCEWA, Ż. P. 1955. The stages of formation of yielding organs in cereals. Moscow University, Moscow.

KUSZELEWSKI, L. 1965. The effect of high doses of mineral fertilizers on light podsolic soils. Rocz. Nauk Roln. Ser A 90: 285–314.

ŁACICOWA, B., and FILIPOWICZ, A. 1972. Research on occurrence of *F. nivale* (Fr.) Ces. on rye field in the Lublin voivodeschip. Ann. Univ. M. Curie-Skłodowska. Sec. E, 27: 211–232.

LANGE, - de la CAMP, M. 1966. Die Wirkungsweise von *Cercosporella herpotrichoides* Fron. dem Erreger der Halmbruchkrankheit des Getreides. II.

LANGER, R. H. 1963. Tillering in herbage grasses. Herb. Abstr. 33:141–148.

LARCHER, W. 1969. Physiological approaches to the measurements of photosynthesis in relation to dry matter production by trees. Photosynthetica 3: 150–166.

LECHNER, L. 1961. Die Fruchtfolge in Getreidesaatgutbau unter besonderer Berücksichtigung der Fusskrankheiten. Saatgutwirtschaft 13: 75–78.

LEVITT, J. 1956. The hardiness of plants. Arci. Press, Inc., New York.

LISTOWSKI, A. 1955. On the influence of autumn and spring drought on the development and yields of winter wheat varieties. Rocz. Nauk Roln. Les. Ser. A 71: 169–195.

LISTOWSKI, A., and DOMAŃSKA, H. 1960. The influence of the autumn and spring drought in the development of winter rye and barley. Rocz. Nauk Roln. Ser. A, 83: 229–241.

MACER, R. C. F., and Van den DRIESSCHE, M. 1966. Yellow rust (*Puccinia striiformis* Westend.) of barley in England, 1960-65. J. Agr. Sci. Camb. 76: 255–265.

MAINS, E. B. 1926. Rye resistant to leaf rust, stem rust and powdery mildew. Agr. Res. 32: 931–972.

MARKOWSKI, A., and DUBERT, F. 1972. Further studies on ribonuclease activity in leaves of winter wheat in the phase of generative and vegetative development. Bull. Acad. Pol. Sci. Ser. Sci. Biol. 20: 339–343.

MARKOWSKI, A., FILEK, W., and MADEJ, M. 1968. DNA and RNA nucleotide composition in winter and spring wheat embryos during germination at 20°C and at vernalization temperature (1°C). Bull. Acad. Pol. Sci. Ser. Sci. Biol. 16: 783–786.

MAZUREK, J., and MAZUREK, J. 1971. Studies on biology of germination in cereals. Pamiet. Puławski 47: 121–129.

MOORE, K. 1963. The influence of climate on a population of tetraploid spring rye. Hereditas 50: 269–305.

MÜNTZING, A. 1954. An analysis of hybrid vigour in tetraploid rye. Hereditas 3: 265–277.

MÜNTZING, A. 1963. Effects of accessory chromosomes in diploid and tetraploid rye. Hereditas 49: 371–426.

NALBORCZYK, T., and NALBORCZYK, E. 1973. A comparative study on productivity of photosynthesis in six spring wheat varieties. Informator o wynikach badań zakończonych w 1971 r. PAN, Warszawa, A6.

OZBUN, J. L., and WALLACE, D. H. 1974. Physiological components of crop yield. Proc. XIX Int Hort. Congr. Warszawa. 1B: 764.

PICHL, J., and RUŽIČKA, B. 1970. Studies on natural occurrence of copper in our essential cereal and pulse crops. Biol. Chem. Vyz. Zviř. 6: 12–14.

PICHLER, F. 1952. Über die Prüfung von Roggensorten auf ihre Anfälligkeit für Schneeschimmel (*Fusarium*). Pflanzenschutzberichte 8: 33–43.

PIELKA, J. 1968. *Fusarium nivale* (Fr.) Ces. on rye crops in Southern Poland. Zesz. Nauk. Wyzsz. Szk. Roln. Krakowie 46: 7–119.

PROCENKO, A. F., and KOLOSHA, O. I. 1969. Physiology of the frost resistant varieties of winter crops. Kiev University, Kiev.

PURVIS, O. N., and GREGORY, F. G. 1952. Studies in vernalization. XII. The reversibility by high temperatures of the vernalized condition in Petkus winter rye. Ann. Bot. N. S. 16:1–21.

RAILO, A. J. 1950. Fungi of the genus *Fusarium*. The Agricultural Publishing House, Moscow.

ROBERTS, B. E., and OSBORNE, D. J. 1973. Protein synthesis and loss of viability in rye embryos. The lability of transferase enzymes during senescence. Biochem. J. 135: 405–410.

ROBERTS, B. E., PAYNE, P. I., and OSBORNE, D. J. 1972. Protein synthesis and the viability of rye grain. Loss of activity of protein synthesizing system *in vitro* associated with loss of viability. Biochem. J. 131: 275–286.

ROEMER, T., FUCHS, W. M., and ISENBEK, K. 1938. Die Züchtung resistender Rassen der Kulturpflanzen. Berlin.

RUEBENBAUER, T. 1964. Genetic foundations of plant breeding. In: Plant Breeding. PWRiL, Warszawa.

RUEBENBAUER, T., and MÜLLER, K. 1963. The research work on tetraploid rye at Borek Fałecki. Biul. IHAR 5-6: 87–90.

RYBAKOVA, M. I., and DENISOVA, R. R. 1972. Changes in the pigment system of winter wheat and rye leaves during autumn and winter. Sel.-choz. Biol. 7: 329–333.

SCHWEIZER, C. J. 1968. Effect of simazine on the protein content of legume and grain crops. Abstr. Meet. Weed Sci., Soc. Am. 30–31.

SŁABOŃSKI, A. 1964. Studies on tetraploid rye. Zesz. Nauk. Wyzsz. Szk. Roln. Szczecinie. 14: 3–17.

SŁABOŃSKI, A., and LIPIŃSKA, D. 1969. A comparison of the influence of water deficiency and excess on yields of diploid and tetraploid rye varieties. Biul. IHAR 3-4: 163–174.

SMETÁNKOVÁ, M., and BAIER, J. 1968. Concentration and nutrient ratio in winter rye with ears different size. Rostl. Výroba. 14: 793–802.

STACHYRA, T. 1959. *Fusarium nivale* in the Carpathians. Prace Naukowe Inst. Ochr. Rośl. 1: 277–308.

STOY, V. 1973. Assimilatbildung und-Verteilung als Komponenten der Ertragsbildung beim Getreide. Angew. Bot. 47: 17–26.

STREBEYKO, P., and DOMAŃSKA, H. 1954. Attempts to determine drought resistance in wheat during germination and emergence. Rocz. Nauk Roln., Ser. A 68: 517–538.

STREBEYKO, D., KRZYWACKA, T., and LEKCZYNSKA, J. 1973. The significance of ear and flag leaf for seed production in some wheat varieties. Informator o wynikach badan' naukowych zakończonych w 1971 r. PAN, Warszawa.

ŚWIEŻYŃSKI, K. 1972. Advance in work on synthesis of initial plant materials for potato breeding in Poland. Postepy Nauk Roln. 19: 95–107.

TARKOWSKI, C. 1966. Cytogenetics and fertility of diploid and tetraploid rye. Postepy Nauk Roln. 13: 35–54.

THOMSON, L. W., and ZALIK, S. 1973. Lipids in rye seedlings in relation to vernalization. Plant Physiol. 52: 268–273.

THORNE, G. N. 1963a. Varietal differences in photosynthesis of ears and leaves of barley. Ann. Bot. N. S. 27: 155–175.

THORNE, G. N. 1963b. Distribution of dry matter between ear and shoot of Plumage Archer and Proctor barley grown in the field. Ann. Bot. N. S. 27: 245–252.

TITOVA, E. N. 1969. The influence of soil properties and fertilizer use on quality of the winter rye grain. Wiest. S-ch. Nauki 14: 139–141.

TYUNOV, A. N., GLUCHICH, K. A., and CHOVKOVA, O. A. 1969. The winter rye. Kolos, Moscow.

VELIKOVSKII, V. 1974. Some methods of rye breeding for resistance to snow mould (*Fusarium nivale*). Eucarpia-Meeting on the breeding of rye, Poznań. VI: 149–152.

WATERHOUSE, S. L. 1953. Experiments in crossing wheat and rye. Proc. Linn. Soc. N. S. W. 78.

WATSON, D. J., THORNE, G. N., and FRENCH, S. A. W. 1958. Physiological causes of differences in grain yield between varieties of barley. Ann. Bot. N. S. 22: 321–351.

CHAPTER 4

MORPHOLOGY AND CHEMISTRY OF THE RYE GRAIN

D. H. SIMMONDS
W. P. CAMPBELL
C.S.I.R.O. Wheat Research Unit
North Ryde, Australia.

I. GENERAL MORPHOLOGY

A. The Rye Inflorescence

The inflorescence of the rye plant *Secale cereale* L. (Figure 1) is a compound, distichous spike whose primary axis, the rachis, bears two opposed rows of lateral spikelets and a single terminal spikelet which is usually rudimentary and sterile. The rachis is a sinuous, flattened, zigzag-shaped structure composed of a number of short nodes and internodes, each internode being narrow at the base and broader at the apex. The spikelets are sessile and arranged alternately in two ranks, the insertion being at the apex of an internode (Figure 1b). The internode is slightly concave on the side next to the spikelet, and the lateral margins of the internodes are fringed with hairs of varying length.

Spikelets consist of a shortened, jointed axis, the rachilla, bearing a variable number of alternately placed, solitary, and sessile florets. Early descriptions of the spikelet usually stated that it was composed of two florets (Hitchcock, 1950), or two florets and a rudimentary third floret (Gates, 1936). More recently, Bonnett (1966) reported that examination of the early stages of spikelet development reveals the presence of six or more florets per spikelet.

At the base of each spikelet there are two oppositely placed, empty, acuminate glumes (Figure 2a, "e"). These are subulate (awl-shaped) structures, about two-thirds the length of the lemmas (le) above them. A scabrous keel extends from the base to the tip of the glume, terminating in a short awn.

Each fertile floret (Figures 2b and 2c) possesses a broad-keeled lemma, terminating in a long scabrid awn (a), and a thin, blunt, two-keeled palea (pa). The floret interior consists of two small lodicules (lo), thick and fleshy at the base but thin, membranous, and ciliate at the apex and margins, together with three stamens and a single glabrous pistil with two feathery stigmas (Figure 2c). The third floret in each spikelet usually consists of a lemma, with a short awn, a palea, and remnants of the stamens. Any florets above the third are represented by shapeless rudiments of the lemmas and paleas enclosed by the lemma of the third floret.

B. Flowering

The first spikelets to anthese (extrude their anthers) are those about one-third

Figure 1. The rye inflorescence: (a) general view just prior to anthesis; (b) enlargement of base of inflorescence just prior to anthesis, showing insertion of spikelets; (c) general view at physiological maturity; and (d) enlargement of inflorescence at physiological maturity, showing insertion of kernels.

below the apex of the spike. Thereafter, flowering proceeds both basipetally and acropetally. Within each spikelet, the basal floret opens first and the others in succession upwards. The whole spike takes approximately four days to complete its flowering, the time being controlled by temperature and humidity.

In each floret, swelling of the lodicules results in the separation of the lemma and palea, followed by spreading of the stigmas and elongation of the filaments of the stamens. At the same time, dehiscence of the pollen sacs takes place, accompanied by pollination of adjacent florets. Individual florets are generally considered to be self-incompatible (Hector, 1936), although fertility exists among the florets on a spike (Obermeyer, 1916). The process of fertilization is

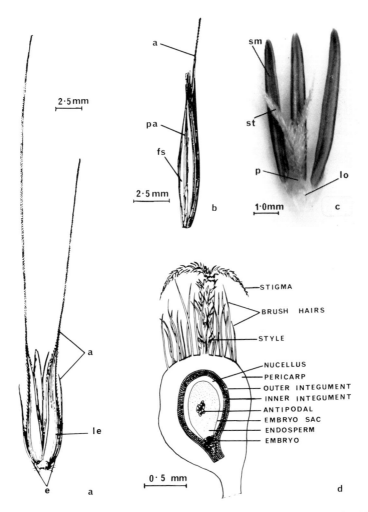

Figure 2. Diagrammatic view of the rye spikelet (a) and floret (b), showing the relationship of floral components (a = awn, e = empty glumes, fs = aborted florets, le = lemma, pa = palea). Taken from Hitchcock (1936) and Bonnett (1966). (c) Dissected floret with lodicules (lo), stamens (s), pistil (p), and stigmas (st). (d) Diagrammatic view of the rye ovary at fertilization.

similar to that in other small-grained cereals. Within 1.5 to 2.5 hr after pollination, the tip of the pollen tube grows down into the stigmatoid tissue in the center of the style, pushing the stigmatic cells apart as it grows between them, eventually passing through the micropyle and into the ovule (Bonnett, 1966). Prior to fertilization, the pollen grain contains three nuclei: the tube nucleus and two male nuclei or gametes. Upon fertilization, the male gametes pass down the tube and are discharged into the embryo sac, which contains an egg, two degenerating synergids, two fused polar nuclei, and several antipodals (Figure 2d).

One of the gametes fuses with the egg to form a zygote or oospore; the second unites with the two polar nuclei, a triple fusion which produces the primary (3n) endosperm nucleus. A few hours after fertilization, the zygote divides transversely into two cells. The larger basal cell forms the suspensor and undergoes little or no further division. The upper cell divides in the transverse and vertical planes, forming a four-celled pro-embryo. Further division and growth in all planes result in a rapidly growing embryo.

Immediately after fertilization and before the zygote starts dividing, the endosperm nucleus commences rapid division, giving rise to numerous free nuclei within the embryo sac (Percival, 1921) and ultimately to the filling of the embryo sac with endosperm tissue.

II. KERNEL STRUCTURE AND DEVELOPMENTAL MORPHOLOGY

A. The Mature Grain

The rye grain is a caryopsis, a small, dry, indehiscent, one-seeded fruit, ranging up to 6–8 mm in length and 2–3 mm in width. The ripe grain is free-threshing and normally grayish yellow in color, although a number of color forms are known. The seed consists of a germ, or embryo, attached through a scutellum to the endosperm and aleurone tissues. These are enclosed by the remnants of the nucellar epidermis, the testa or seed coat, and the pericarp or fruit coat. The latter surrounds the whole seed and adheres closely to it (Figures 3a and 3b).

The crease, or furrow, extends the full length of the grain on the ventral side. On the dorsal side, the outline of the embryo can be seen at the base. Mature kernels often have a shrivelled appearance. The developing kernel is served by four vascular traces emanating from a basal vascular bundle. Three of these traces extend through the pericarp, one dorsal and two lateral to the embryo sac. The fourth passes up the crease of the developing kernel on the ventral side, and its remnants may be observed in this area at maturity.

B. The Developing Grain

The following description of developing grain morphology is based on material (*S. cereale* L., cv. Florida Black) grown in the glasshouse under rather cool seasonal conditions. Maturation of the kernels took between 55–65 days, considerably longer than 42–45 days usually encountered with material field-grown in the correct season. Each stage of development (Figure 4) was, however, normal in its gross morphology, and its microscopic structure corresponded exactly to that of other material grown under more normal conditions. The times between stages have therefore been adjusted to correspond with the more normal maturation time of 45 days. Collection of specimens, fixation, embedding, and sectioning followed usual procedures (Feder and O'Brien, 1968; Spurr, 1969).

The embryo is enclosed by several tissue layers which during kernel development undergo marked and predictable changes. The pericarp, which is 8 to 16 cells thick, constitutes the outermost zone of the kernel. Adjacent are the outer and inner integuments, each composed of two cell layers, and finally the nucellar epidermis, the outermost layer of the nucellus. The latter comprises the ovule, within which the embryo and endosperm develop. A photomicrograph of these cell layers is illustrated in Figure 5a. The changes which they undergo as the grain develops and matures are continous and interrelated; for convenience, the morphogenesis of each will be described separately.

PERICARP OR FRUIT COAT

The pericarp, which surrounds the entire grain and completely encloses the embryo sac, is maternal in origin, being the original wall of the ovary. Initially at anthesis, it consists of a layer, 8 to 16 cells thick, bounded both externally and

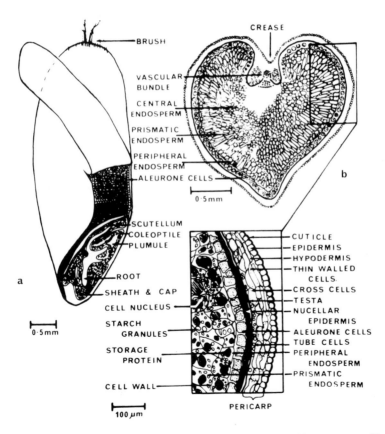

Figure 3. Diagrammatic view of rye grain in longitudinal section (a) and in transverse mid-section (b).

internally by an epidermis. Immediately below the outer epidermis lies the hypodermis, and adjacent to this is a delicate middle parenchyma composed of numerous spherical, thin-walled cells (Figure 5a).

In the first few days following anthesis, the tissues constituting the pericarp continue to enlarge, with the active deposition of small starch granules (Figure 5b).

Degeneration of the pericarp commences approximately 3-4 days after anthesis, with cell lysis commencing in the middle parenchyma, two layers removed from the inner epidermis. The first indication of impending destruction is a breakdown in the regular array of cells due to dissolution of their middle lamellae (Figure 5c), and is similar to that described by Simmonds (1974) and by Dedio *et al.* (1974) in triticale.

By 7 days post-anthesis, cell walls are extensively degraded, and cell contents, including the small spherical starch granules, have been released into the resulting empty space between the inner and outer epidermis (Figure 5d). This space is, however, quickly filled by expansion of the endosperm and associated tissues. By approximately 14 days after anthesis, only two or three layers of parenchyma tissue remain undisturbed (Figure 5e), and by 24 days post-anthesis, only the outer epidermis and hypodermis remain (Figure 5g).

The three vascular traces originally present in the pericarp are destroyed,

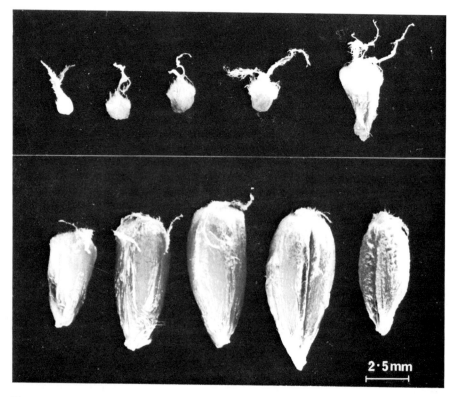

Figure 4. Gross morphology of rye kernels at different stages of development.

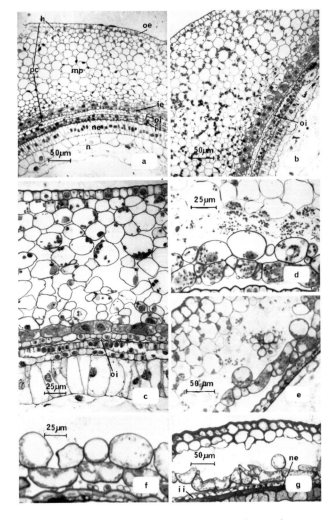

Figure 5. Photomicrographs showing stages in the degeneration of the pericarp, testa, and nucellar tissues in developing rye kernel: (a) Mid-transverse section at anthesis showing pericarp (pc), consisting of outer epidermis (oe) and inner epidermic (ie), hypodermis (h), and thin-walled middle parenchyma (mp); inner and outer integuments (ii and oi, respectively), each composed of two cell layers; nucellar epidermis (ne), the outermost layer of the nucellus (n). (b) Rye kernel 2–3 days post-anthesis showing active deposition of small starch granules (s) in the middle parenchyma and the initial stage in the collapse of the outer integument (oi). (c) Section 3–4 days post-anthesis showing initial stages in breakdown of the pericarp. At this stage, degeneration of the outer integument (oi) is almost complete. (d) Further stage in the destruction of the middle parenchyma, 7 days post-anthesis. Note extensive degradation of cell walls and the release of small starch granules. (e) Rye kernel 14 days post-anthesis when only two or three layers of the parenchyma remain undisturbed and the nucellar epidermis has commenced to degenerate. (f) Section 18 days post-anthesis showing changes in the two innermost layers of the pericarp. (g) At 24 days post-anthesis only the outer epidermis and hypodermis of the pericarp are intact. Note also the remains of the inner integument (ii) and nucellar epidermis (ne).

although they persist until approximately 18 days post-anthesis (Figure 5e). The fourth trace, which extends through the funiculus to serve the developing ovule, remains intact at maturity.

The mechanisms responsible for the destruction of the pericarp tissues are not known, but are presumably enzymatic. However, Dedio *et al.* (1974) have shown that dissolution of the starch granules in the pericarp of triticale is effected by an α-amylase specifically released in this tissue between 5 and 22 days after anthesis (see Section IV-C).

The two innermost layers of the pericarp (Figure 5) do not degenerate with the cells of the middle parenchyma, although they change markedly in appearance as the grain matures. The layer adjacent to the inner epidermis consists initially of large cylindrical cells, many times longer than they are wide, with their long axes at right angles to the long axis of the grain. Until some 10 days post-anthesis, these cells are photosynthetically active since they develop, soon after anthesis, numbers of chloroplasts and associated grana. These structures are clearly seen under the electron microscope (Figure 6) which also shows the intact cells of the nucellar epidermis at this stage.

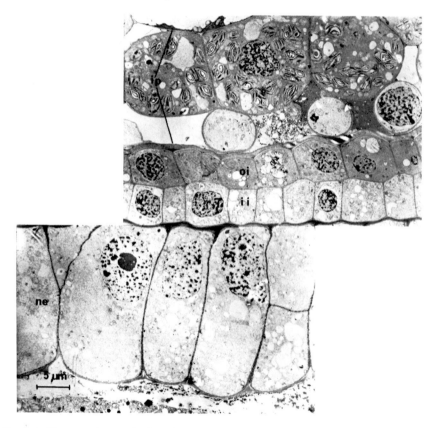

Figure 6. Electron micrograph of two innermost layers of the pericarp (ip), integuments (ii and oi), and nucellar epidermal cells (ne) taken 7 days post-anthesis. This figure is a montage of two adjacent fields.

As the grain matures, these cells enlarge, their walls thicken slightly, and their cytoplasmic contents degenerate (Figure 5f). At maturity, they form a disorganized layer of empty cells whose long axes run transversely. At this stage they are known as the "cross" cells.

At anthesis, the inner epidermis of the pericarp consists of a layer of somewhat elongated cells parallel to the outer epidermis and at right angles to the cross cell layer (Figure 5a). By 8 days after anthesis their walls have thickened slightly, and individual cells have become separated from one another (Figure 6). They lie with their long axes parallel with the long axis of the grain and in the mature kernel are referred to as the "tube" cells (Figure 9b). As the kernel dries out, the tube cells are crushed into close contact with the testa, with the result that they are sometimes difficult to discern in the mature grain.

TESTA OR SEED COAT

At anthesis, the embryo sac is enclosed by two integuments, inner and outer, each composed of two cell layers (Figure 5a). A thin cuticle is evident on the surface of the inner integument. However, the cells of the outer integument collapse quite rapidly (Figure 5b), and by 4–5 days after anthesis, they are no longer visible in transverse sections (Figure 5c). A similar situation exists in triticale, where the large oval and tubular cells give rise to the cross and tube cells at maturity and therefore comprise the inner pericarp rather than the outer integument as suggested by Simmonds (1974). Cells of the inner integument remain as a visible layer until maturity. Their cuticle becomes extensively thickened, and the cell contents start to degenerate by 7–10 days after anthesis. By 18 days post-anthesis, the walls have collapsed, and the contents have been compressed by the enlarging endosperm (Figure 5g). The crushed and desiccated remains of the inner integument, together with its thickened cuticle, constitute the testa or seed coat. In the mature grain, it may be recognized as a thin, strongly staining line lying between the pericarp and the nucellar epidermis. Staining with toluidine blue is probably due to its content of polyphenols. In addition, the cuticle stains strongly with lipid stains such as Sudan IV. Together with the pigment strand, the seed coat forms a waxy, water-repellent zone completely surrounding the endosperm and embryo. The sequence of events in the development and destruction of the pericarp and integuments (both inner and outer) is thus essentially as reported by Percival (1921) and Peterson (1965) for wheat and by Hector (1936) for rye.

NUCELLAR EPIDERMIS

The nucellar epidermis, otherwise known as the hyaline layer, lies between the testa and the aleurone and is closely united to both. In the developing kernel it forms the outermost layer of the nucellus (Figure 5a), the tissue comprising the ovule, within which the embryo and endosperm develop. This tissue also serves as a source of nutrients in the early stages for the developing embryo and endosperm. At 2–3 days post-anthesis, several layers of intact nucellar tissue may be seen (Figure 5b), but by 6–7 days, only the nucellar epidermis remains (Figures 5c and 6). At this stage it may be recognized as a band of columnar cells with their long axes perpendicular to the long axis of the grain and arranged radially around the developing endosperm (Figure 7a). Shortly thereafter its cytoplasm degenerates, the radial walls break down, and the cells collapse as the endosperm enlarges (Figures 5e and 5g). In the crease region, the cell remnants are less

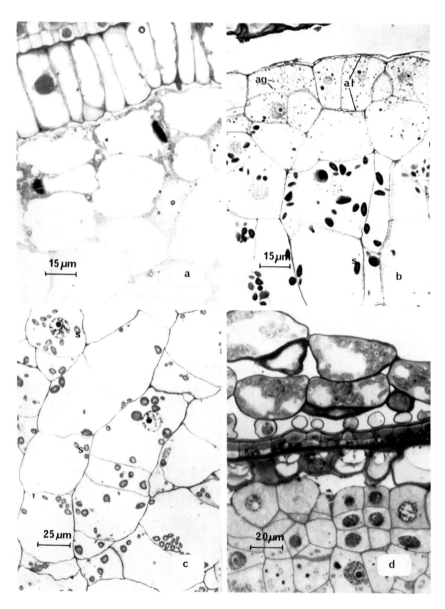

Figure 7. Photomicrographs showing stages in the development of the endosperm and aleurone in the maturing kernel. (a) Formation of cell walls 24–48 hr post-anthesis following free nuclear division. (b) Section 10–12 days post-anthesis showing initial stages in the differentiation of aleurone (al) with the formation of aleurone granules (ag). Note also the deposition of starch (s) in the peripheral cytoplasm of starchy endosperm cells. (c) Deposition of starch granules (s) in the peripheral cytoplasm of central endosperm cells at 10–12 days post-anthesis. (d) Endosperm 18 days post-anthesis showing radial columns of cells resulting from periclinal divisions of the peripheral layer.

compressed, and cavities are occasionally visible. At maturity it represents the innermost layer derived from the maternal parent, and as in triticale (Simmonds, 1974) and wheat (Fairclough, 1947, cited by Bradbury *et al.,* 1956), it appears that the layer surrounds the entire grain except for the greater part of the embryo. It is generally hyaline in appearance, but occasionally dark inclusions can be observed. These have been derived from the contents of the original nucellar epidermal cells.

ENDOSPERM AND ALEURONE

Both the aleurone and the starchy endosperm cells are formed by divisions of a peripheral meristematic layer, just below the nucellar epidermis. The aleurone is botanically the outer layer of the endosperm and in rye is generally one cell thick. It completely surrounds the starchy endosperm and merges into the scutellum located between the endosperm and embryo. In surface view, the cells appear polygonal, while in transverse section, the cells are square to slightly rectangular with elongation normally perpendicular to the long axis of the grain in the radial direction.

In the mature grain, the aleurone is characterized by the presence of numerous, intensely staining aleurone granules. Clustering around each granule are large numbers of lipid droplets, called spherosomes, which comprise the lipid storage reserves of the grain. Externally the layer is tightly pressed against the remains of the nucellar epidermis, while internally it is joined without discontinuity to the endosperm (Figure 9a).

The starchy endosperm represents the bulk of the kernel. It is composed of three types of cells: peripheral or sub-aleurone, prismatic, and central (MacMasters *et al.,* 1971) which differ in shape, size, and location within the kernel. The different cell types arise as a result of the distribution and timing of cell division during early kernel development (Figures 9b and 9c). The sub-aleurone cells develop towards the end of endosperm differentiation and comprise one to three layers of roughly isodiametric cells. The radially elongated prismatic cells constitute the bulk of the endosperm tissue and extend from the dorsal area of the kernel to the area surrounding the crease. The central endosperm cells are located around the head of the crease and extend towards the center of each cheek. Both in wheat and rye they are irregular in size and shape, having been formed at a very early stage of endosperm development.

The fusion of one of the two male gametes with the two polar nuclei at fertilization has already been described. The events immediately following this are still a matter for discussion. Brenchley (1909) claimed that coenocytic development, or free nuclear division, was responsible for the formation of the endosperm in its entirety. McLennon (1920) and Gordon (1922) believed that all endosperm cells arose by serial tangential divisions of a cambial layer lining the embryo sac, this layer being derived initially by multiple divisions of the first endosperm nucleus. Other workers (Sandstedt, 1946; Yampolsky, 1957; Jennings and Morton, 1963; Buttrose, 1963) have combined the two theories, suggesting that the inner endosperm arises as described by Brenchley and the outer layers as described by Gordon. Evers (1970), in a very thorough investigation, found that in wheat there was no evidence to support either the coenocytic development theory of Brenchley or the combined theory. Evers considered that Gordon's observations accurately described the situation as it exists in wheat. However, he also stated: "If a free nuclear stage exists in the development of endosperm from the first endosperm nucleus, cellularization of the syncytium must result in the formation of the meristematic tissue, and not the storage tissue of the starchy endosperm." Our observations both in rye and triticale (Shealy and Simmonds, 1973; Simmonds, 1974) indicate that rapid, free nuclear division, normally in a synchronized manner, takes place for a period of approximately 72 hr after fertilization. This gives rise to the peripheral, meristematic zone of the endosperm. Subsequently, the rate of nuclear division decreases markedly, and cell walls are laid down, both in the peripheral and mid-endosperm areas, within the following 24-48 hr (Figure 7a); following this, further cells arise by division of the peripheral cambial zone, both periclinally (tangentially) and anticlinally (radially).

Cytokinesis in rye, first evident some 5-6 days after anthesis, results in the production of radial columns of cells (Figures 7d and 9c), although this arrangement is frequently disturbed in the cheek regions. As already indicated, peripheral cells around the top of the crease lose their meristematic capacity early in kernel development, resulting in a considerable disorganization of cell arrangement in this area and the production of so-called central endosperm cells.

In rye, cell division to yield starchy endosperm continues until approximately 18 days post-anthesis. Subsequent to this stage, the outer cell layer develops considerable quantities of aleurone granules, and the walls become distinctly thickened (Figure 9a). Under the light microscope, the first indication of differentiation of aleurone cells may be detected by 10-12 days post-anthesis (Figure 7b) when the peripheral endosperm layer may be seen to contain both starch granules and the first initials of aleurone granules. In several sections, occasional aleurone granules have been noted in cells one or even two layers below the peripheral layer. Their presence in these layers is probably due to continued division of the peripheral layer after the granules are first differentiated. However, in the deeper layers, the granules are resorbed and their contents presumably used as an additional source of energy for the developing tissue.

In some areas, the normally regular arrangement of cells in the peripheral layer may be disrupted by an invagination into the underlying starchy endosperm tissue (Figure 9d). Similar observations have been made in triticale (Simmonds, 1974), and the phenomenon is considered to be responsible for grain shrivelling in this cereal. In the case of triticale, the invaginations appeared to be caused both by a thickening and intrusion of the nucellar epidermis between adjacent dividing cells and by degradation of the meristematic layer, possibly by greater than normal cytolytic activity at the expanding edge of the developing endosperm. Such cytolytic enzymes would normally be directed towards the destruction of nucellar tissue. Neither of these phenomena was specifically observed in the rye samples examined. However, in the regions where the intrusions did occur, the space between the peripheral layer and the nucellar epidermis was partly filled with a granular material, possibly derived from the destruction of other nucellar tissue.

Starch deposition in the developing endosperm is first noticeable under the light microscope at about 10 days post-anthesis and under the electron microscope at about 7 days. The granules generally appear first in the peripheral cytoplasm, adjacent to the cell walls (Figures 7b and 7c). As the grain develops, existing granules enlarge, and new granules are initiated. Deposition does not appear to start in any particular region. However, in the mature grain, prismatic and central endosperm cells have considerably greater numbers of starch granules distributed throughout their cytoplasm than in the sub-aleurone layer, where they are restricted to the peripheral regions (Figures 9b and 9c). Starch granules are generally of two principal size distributions: large lenticular granules up to about 35 μm in the longest dimension and smaller spherical granules with diameters less than 10 μm. Both classes are observed in the prismatic and central endosperm cells, but granules in the sub-aleurone layers are usually intermediate in size. The distribution of starch granule size in rye is dealt with further in Section V.

The synthesis and deposition of storage protein first occur at approximately 12–14 days after anthesis when small areas can be discerned in the cytoplasm of the cells of the sub-aleurone layers. In some cases, as shown in Figure 8, the developing protein deposit is associated with a roughly spherical area (pb) approximately 3–4 μm in diameter. In other cases, the protein deposits (pd) are seen to be free in the cytoplasm. Under the light microscope, the deposits are seen to accumulate within a vacuole, where they eventually coalesce (Figure 9e) to form a protein "pool" (Campbell *et al.*, 1974). Further protein synthesis has been reported to take place in association with areas of rough endoplasmic reticulum (Barlow *et al.*, 1973). Protein synthesis continues until grain dehydration and accompanying disruption of cellular structure occur just prior to maturity.

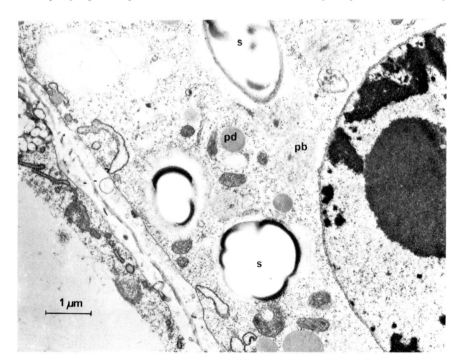

Figure 8. Electron micrograph of sub-aleurone cells at 12–14 days post-anthesis showing developing protein bodies (pb), cytoplasmic protein deposits (pd), and starch granules (s).

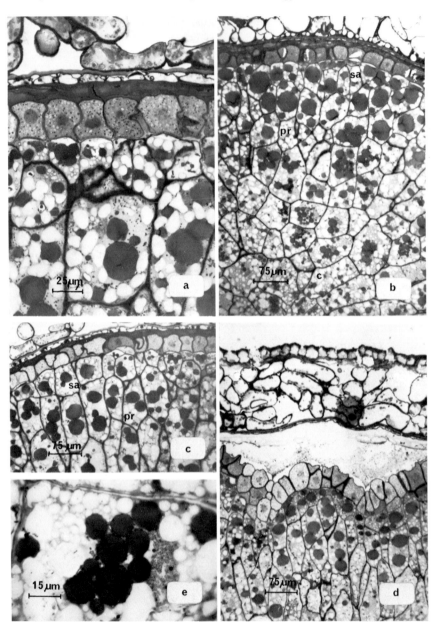

Figure 9. Photomicrographs showing stages in the development of the endosperm and aleurone. (a) Rye kernel at 24 days post-anthesis showing aleurone and sub-aleurone cells. (b) and (c) At 30 days post-anthesis showing peripheral or sub-aleurone (sa), prismatic (pr), and central (c) endosperm cells. Note the gradient of protein deposition in moving across the kernel, and the radial columns of cells. (d) Invagination of aleurone into starchy endosperm tissue at 30 days post-anthesis. (e) Prismatic endosperm 34 days post-anthesis showing coalescence of vacuolar protein deposits to form a protein "pool."

However, the deposition is not uniform throughout the endosperm—there is a distinct drop in concentration from the periphery to the center of the kernel (Figure 9b). Evers (1970) has calculated that in the mature wheat kernel, all cells contain approximately the same weight of storage protein regardless of their position in the endosperm. Insufficient evidence is available to determine whether the condition is similar in rye. However, in some cases, initiation of protein synthesis in central and prismatic endosperm cells was observed to be delayed up to 5 days after it was well established in the peripheral cells, suggesting that the observed gradient in protein concentration may not be entirely due to dilution of mid-endosperm cell contents by starch.

THE EMBRYO

The embryo of rye closely resembles that of wheat (Hector, 1936). It lies at the base of the kernel on the dorsal side and consists of a short axis with the plumule at its apex and the primary root at its base. The scutellum, a shield-shaped structure, lies at the back of the embryo, appressed to the endosperm, and functions initially as a food reserve for the embryo on germination, and later, for the digestion and absorption of endosperm reserves. At the front of the embryo, and opposite the scutellum, is a small projection known as the epiblast. Between the scutellum and the epiblast then, lies a young plant "in embryo," but already differentiated into rudimentary roots, stems, and leaves.

III. GENERAL COMPOSITION OF THE MATURE RYE KERNEL

A. Proximate Analysis

Among cereal crops, rye alone resembles wheat in yielding a flour which, when kneaded with water, yields a cohesive, extensible dough. This property is also possessed by the wheat-rye hybrid, triticale, and is considered to be a function of the storage protein fraction in each of these three cereals. It must be noted, however, that both rye and triticale yield doughs that are considerably inferior to wheat doughs in their rheological properties and suitability for breadmaking. Characteristic of cereal storage proteins is the high proportion of glutamic acid and proline and low proportions of lysine, methionine, and tryptophan. As will be shown in greater detail later, rye differs from wheat, barley, and most other cereals in having a comparatively higher proportion of water- and salt-soluble protein, both of which have an improved content of the essential amino acid lysine. Rye protein as a whole is therefore considered to be superior to wheat and most other cereal proteins in biological value. However, its availability is severely reduced by factors which restrict its true digestibility when presented as rye flour in the diet (Eggum, 1968).

Table I presents comparative figures for the proximate composition of rye, several cereals, and soybean. It should be stressed that the values given in Table I represent average or typical values only.

Protein contents ranging between 6.5 and 14.5% have been reported for rye, the higher values usually being obtained under North American conditions of cultivation where nitrogen fertilizer applications are generally higher (Kent-Jones and Amos, 1967). Considerable disagreement exists as to the correct factor

TABLE 1

Proximate composition of rye compared with other grains and soybean
(moisture free basis)

	Rye[a] %	Triticale[b] %	Wheat[a] %	Barley[a]		Corn[a] %	Oats[c]		Rice[a]		Soy[d] %
				Whole Grain %	Kernel Only %		Whole Grain %	Kernel Only %	Whole Grain %	Kernel Only %	
Protein	13.4	14.8	14.3	13.1	14.5	10.4	13.0	17.0	8.2	9.4	42.8
Ether extract	1.8	1.5	1.9	2.1	2.1	4.5	5.5	7.7	2.2	1.8	19.6
Crude fiber	2.6	3.1	2.9	6.0	2.1	2.4	11.8	1.6	10.1	0.9	5.5
Ash	2.1	2.0	2.0	3.1	2.3	1.5	3.7	2.0	5.7	1.1	5.0
Nitrogen free extract[e]	80.1	78.6	78.9	75.7	79.0	81.2	66.0	71.6	73.8	86.8	27.1

[a] Miller (1958).
[b] Munck (1972).
[c] Kent (1966).
[d] Cartter and Hopper (1942).
[e] An estimate of total carbohydrate.

to be used to convert Kjeldahl nitrogen to total protein in the case of rye. Figures in the literature range from 5.64 to 6.25, the former, due to Tkachuk (1969), being based on a very thorough examination of the amino acid composition of total rye protein (Tkachuk and Irvine, 1969). The World Health Organization (1973) has recently adopted the figure of 5.83 for both rye wholemeal and flour. For convenience in comparisons between wheat and rye, however, the conversion figure of 5.7 will be used throughout this chapter. This figure is also a reasonable compromise between the two figures 5.64 and 5.83 quoted above.

The principal component of the nitrogen-free extract is starch, the proportion of which varies inversely with protein content. The figures quoted in Table I are thus average values, with fluctuations of several percent occurring under varying environmental conditions. The figures quoted for crude fiber provide an estimate of the nutritionally unavailable proportion of carbohydrate in the cereal. The limitations of assay methods for crude fiber have been discussed by van Soest (1966), Southgate (1969), and Munck (1972). The presence of lignin is particularly deleterious to the availability of carbohydrates and protein in all animals.

B. Minerals

The composition of the ash from the rye kernel is similar to that of other cereals as summarized in Table II.

Ash content is particularly high in the aleurone layer of rye, due primarily to the presence in these cells of aleurone bodies, which consist of phytin granules, a mixture of the potassium and magnesium salts of myoinositol hexaphosphate (Jacobsen *et al.,* 1971; Salmon, 1940; Pomeranz, 1973) surrounded by a protein-containing envelope (Lui and Altschul, 1967). Other portions of the kernel contributing to the ash content are the outer pericarp and the aleurone and endosperm cell walls. X-Ray analysis (Pomeranz, 1973) indicates that both iron and calcium are present in very low concentrations in the aleurone layer.

C. Vitamins and Anti-Nutritional Factors

VITAMINS

Rye, in common with other cereals, is an important source of vitamins such as thiamin, nicotinic acid, riboflavin, pyridoxine, pantothenic acid, and tocopherol. These components are mainly present in the embryo, scutellum, and aleurone layer of the seed. A comparison with other cereals is given in Table III.

The distribution of these substances within the various tissues comprising the rye kernel has been studied in the case of thiamin. Table IV shows the distribution of this vitamin between the embryo and the scutellum in a variety of cereals.

ANTI-NUTRITIONAL FACTORS

The feeding of rye to animals, particularly chickens, in large amounts has been considered disadvantageous for many years (Knieriem, 1900; MacAuliffe and McGinnis, 1971; Munck, 1972). Where the food is ingested in a fairly dry state, as in the case of chickens, the high viscosity of rye mixtures may play an important role in this problem (Munck, 1968). However, rye is known to contain significant quantities of substances having deleterious physiological effects on other

TABLE II

Mineral composition of rye compared with other grains and soybean

(mg/100 g, dry basis)

	Rye[a]	Wheat[a]	Barley[a]		Corn[a]	Oats[b]		Rice[b]		Soy[c]
			Whole Grain	Kernel Only		Whole Grain	Kernel Only	Whole Grain	Kernel Only	
Phosphorus	380	410	470	400	310	340	400	285	290	660
Potassium	520	580	630	600	330	460	380	340	120	1670
Calcium	70	60	90	80	30	95	66	68	67	280
Magnesium	130	180	140	130	140	140	120	90	47	220
Iron	9	6	6	...	2	7	4	...	6	10
Copper	0.9	0.8	0.9	...	0.2	4	5	0.3	0.4	1.2
Manganese	7.5	5.5	1.8	...	0.6	5	4	6	2	2.8

[a] Miller (1958).
[b] Kent (1966).
[c] Markley (1950).

TABLE III

Composition of B group vitamins and carotene in rye compared with other grains and soybean

(mg/100 g, dry basis)

	Rye[a]	Wheat[a]	Barley[a] (Whole)	Corn[a]	Oats[a] (Whole)	Rice[a] (Brown)	Soy[b]
Thiamine	0.44	0.55	0.57	0.44	0.70	0.33	0.90
Riboflavin	0.18	0.13	0.22	0.13	0.18	0.09	0.23
Niacin	1.5	6.4	6.4	2.6	1.8	4.9	2.0
Pantothenic acid	0.77	1.36	0.73	0.70	1.4	1.2	1.2
Pyridoxine	0.33	0.53	0.33	0.57	0.13	0.79	0.64
Carotene	0	0	0.04	0.40	0	0.013	2.20

[a] Miller (1958).
[b] Cartter and Hopper (1942).

organisms. Virtanen (1964) has reviewed the presence of benzoxazolinones, inhibitors to snow mold *(Fusarium nivale)*, which are abundant in rye and wheat endosperms. Rye is particularly rich in 5-alkyl resorcinols (Wieringa, 1967; Munck, 1969) which are toxic to animals. According to Wieringa (1967), rye seeds may contain as much as 0.5% by weight of alkyl resorcinols, more than twice the amount present in wheat. Table V shows the relative amounts of these compounds present in rye, wheat, and triticale.

Wieringa (1967) found considerable variation in alkyl resorcinol content within a single sample of rye. Heavier kernels, containing a higher proportion of endosperm, contained relatively less alkyl resorcinols per grain than lighter kernels. Chromatographic examination confirmed that these toxic factors were localized in the pericarp and did not occur in the endosperm or germ.

Zillman (1973, unpublished data quoted by Hulse and Laing, 1974), obtained results suggesting that the alkyl resorcinol content of the grain *per se* is not detrimental either to growth rate or food intake of mice and that contrary to certain previous reports, the feeding value of both rye and triticale were superior to wheat. The importance of ensuring the absence of ergot and other mold mycotoxins in feeding experiments should be stressed in this connection.

Phytic acid (myoinositol hexaphosphoric acid) and its salts occur, as mentioned above, in the aleurone cells of rye and other cereals. This compound acts as an anti-nutritional factor because of its ability to form highly insoluble salts with mineral elements including calcium, iron, magnesium, and zinc. It has been reported (Hay, 1942; Kent-Jones and Amos, 1967) that in commercial wheat products, phytic acid content runs parallel to fiber content. Both phytic acid and its salts are hydrolyzed by the enzyme phytase to yield inositol and

TABLE IV
Distribution of thiamin between embryo and scutellum in rye compared with other grains

Cereal Grain	Proportion of Tissue in Kernel		Proportion of Total Thiamin in Tissue	
	Embryo %	Scutellum %	Embryo %	Scutellum %
Rye	1.8	1.73	5	82
Barley	1.85	1.53	8	49
Corn	1.15	7.25	8	85
Oats	1.6	2.13	4.5	28
Rice	1.0	1.25	11	47
Wheat	1.2	1.54	3	59

TABLE V
Relative amounts of alkyl resorcinols present in rye, wheat, and triticale[a]

Cereal	Number of Varieties	Alkyl Resorcinols (units)
Rye (mean and range)	15	161 (129–192)
Wheat (mean and range)	18	69 (56–79)
Triticale (mean and range)	19	97 (78–124)

[a]Source: Munck (1972).

phosphoric acid. Wheat and rye are rich in phytase, whose distribution throughout the grain parallels that of its substrate. The phytase of rye has been reported to be more active than that of wheat (Gontzea and Sutzescu, 1968).

Anti-trypsin factors have been reported by several authors (Laporte and Tremolieres, 1962; Polanowski, 1967; Chen, 1973, unpublished data reported by Hulse and Laing, 1974) to be present in wheat and rye. The factor in rye was found by Polanowski (1967) to be located only in the endosperm and not in the germ, seed coat, or aleurone layer. It is comparatively stable to heating, although various authors differ on this aspect. As determined by gel filtration, its molecular weight was reported by Madl and Tsen (1973b) to be 12,000–14,000. It was thermally stable at 100°C for 1 hr. The presence of a chymotrypsin inhibitor in rye, triticale, and wheat was also reported by Madl and Tsen (1974). This substance had a molecular weight of 17,000–19,000 and was inactivated by heating for 10 min at 70°C. Both factors yield several protein bands when subjected to polyacrylamide gel electrophoresis.

Further details of toxic factors in rye, wheat, and triticale are discussed by Hulse and Laing (1974).

D. Amino Acids

As already stated, the amino acid composition of rye protein is claimed to be nutritionally superior to that of most other cereals. This is due to the higher proportion of lysine present in the water-soluble (albumin and globulin) proteins and the higher proportion of these proteins present, compared with the storage group (Chen and Bushuk, 1970a).

Table VI compares the amino acid composition of whole rye meal with that of wheat and triticale, showing both the mean and ranges encountered. The figures are those collated by the Food and Agriculture Organization (1970).

The higher lysine content in rye, compared with both wheat and triticale meal, is apparent. Bran and germ fractions contain a higher proportion again of this essential amino acid. Since the nutritional quality of a foodstuff is dependent upon the quantity of the first limiting amino acid present, the concept of a chemical score has been developed, initially by Mitchell and Block (1946) and later by a joint FAO/WHO Expert Committee (World Health Organization, 1973). The quantity of a particular amino acid present in the foodstuff is expressed as a percentage of the amount present in a standard reference pattern. This pattern has been arrived at as a compromise based on the amino acid

TABLE VI
Average amino acid content of rye, triticale, and wheat as determined by ion-exchange chromatography[a] (g of amino acid/100 g of total nitrogen; ranges encountered in the literature in parentheses)

Amino Acid	Rye Whole Meal	Triticale[b] (1972-73)	Wheat Meal
Isoleucine	21.9 (20.0–24.2)	18.7	20.4 (18.8–21.4)
Leucine	38.5 (36.1–40.6)	45.0	41.7 (37.1–45.0)

Table VI *(continued)*

Amino Acid	Rye Whole Meal	Triticale[b] (1972-73)	Wheat Meal
Lysine	21.2 (15.1–28.1)	19.6	17.9 (13.1–24.9)
Methionine	9.1 (5.9–18.1)	6.0	9.4 (6.3–15.6)
Cystine	11.9 (8.5–15.6)	7.9	15.9 (11.1–21.2)
Phenylalanine	27.6 (25.0–30.0)	28.6	28.2 (23.4–33.8)
Tyrosine	12.0 (7.6–17.5)	19.5	18.7 (8.6–22.5)
Threonine	20.9 (19.1–23.1)	19.6	18.3 (14.8–22.2)
Tryptophan	4.6 (3.4–8.8)	6.3	6.8 (5.1–13.6)
Valine	29.7 (20.6–34.3)	24.2	27.6 (22.8–32.5)
Arginine	28.6 (18.4–34.4)	38.2	28.8 (23.4–34.4)
Histidine	13.8 (12.5–16.5)	13.3	14.3 (12.5–16.3)
Alanine	26.6 (23.5–30.2)	25.8	22.6 (18.8–31.4)
Aspartic acid	44.7 (38.3–51.1)	41.6	30.8 (26.3–33.8)
Glutamic acid	151.1 (135.6–167.6)	152.8	186.6 (158.1–201.9)
Glycine	27.1 (25.0–30.2)	26.5	25.4 (18.8–27.5)
Proline	58.6 (51.7–73.8)	52.1	62.1 (55.0–73.6)
Serine	27.0 (25.0–30.6)	25.0	28.7 (25.6–31.9)
First limiting amino acid	Tryptophan		Lysine
Second limiting amino acid	Isoleucine		Isoleucine

[a]Source: Food and Agriculture Organization (1970) via Hulse and Laing (1974).
[b]Average of data for three advanced triticales produced at CIMMYT, Mexico, in 1972-73.

TABLE VII

Chemical scores for rye and other cereals compared with soybean and the World Health Organization Standard Reference Pattern (the data presented in TABLE VI have been used as a basis for calculation)[a]

Amino Acid	WHO (1973) Standard Reference (mg/g protein)	Rye Amino Acid Content (mg/g protein)	Rye Chemical Score	Wheat Amino Acid Content (mg/g protein)	Wheat Chemical Score	Maize Amino Acid Content (mg/g protein)	Maize Chemical Score	Soybean Amino Acid Content (mg/g protein)	Soybean Chemical Score
Lysine	55	37	67	31	56	27	49	70	127
Threonine	40	37	92	31	77	36	90	42	105
Methionine + cystine	35	37	105	43	123	35	100	28	80
Leucine	70	67	96	72	103	125	180	85	121
Isoleucine	40	39	97	35	88	37	93	50	125
Valine	50	52	104	47	94	48	96	53	106
Phenylalanine + tyrosine	60	69	115	81	135	87	145	89	148
Tryptophan	10	8	80	...	...	7	70	14	140

[a]Source: Modified from Hulse and Laing (1974).

patterns of egg and human milk protein and upon estimated amino acid requirements for humans. A summary of the chemical score of whole rye meal compared with this reference pattern, and also with other cereals and soybean, is given in Table VII.

These figures show that in spite of its higher lysine content, this amino acid is still likely to be the first nutritionally limiting amino acid in whole rye meal, with tryptophan as the second. This has been confirmed by the work of Kies and Fox (1970a,b) and Kihlberg and Ericson (1964). On the other hand, Janicki and Kowalczyk (1965) have reported that methionine and isoleucine were limiting amino acids in rye; and the Food and Agriculture Organization (1970), as shown in Table VI, has claimed that tryptophan and isoleucine are the first and second limiting amino acids, respectively. It would appear that the matter requires further clarification. However, the improved nutritional status of rye compared with wheat, because of its relatively higher lysine content, has been well documented by several workers (Mitchell and Hamilton, 1929; Kon and Markuze, 1931; Johnson and Palmer, 1934; Jones *et al.,* 1948; Sure, 1954; de Vuyst *et al.,* 1958; Bixler *et al.,* 1968; Knipfel, 1969; Kies and Fox, 1970a,b; McGinnis, 1972).

The amino acid composition of several Polish rye varieties has been reported by Janicki and Kowalczyk (1965). In Table VIII their results are compared with those obtained by Ewart (1967) for English rye and Tkachuk and Irvine (1969) for Canadian rye.

TABLE VIII
Amino acid composition of Polish, English, and Canadian rye varieties
(g amino acid / 100 g whole rye protein)

	Polish Rye[a] cv. Rogolinskie	English Rye[b]	Canadian Rye[c]
Protein (N × 5.7)	9.27	7.24	...
Alanine	4.91	4.54	4.07
Ammonia	...	2.72	3.19
Arginine	6.12	4.93	4.60
Aspartic acid	7.84	6.46	7.09
Cystine/Cysteine	...	3.45	2.51
Glycine	4.90	3.99	3.98
Glutamic acid	23.81	33.46	30.18
Histidine	2.66	2.38	2.30
Isoleucine	3.72	4.21	4.00
Leucine	6.72	7.79	6.56
Lysine	4.54	3.72	3.18
Methionine	1.93	1.93	1.28
Phenylalanine	3.56	5.53	4.91
Proline	9.45	14.23	11.4
Serine	4.45	6.12	4.70
Threonine	3.54	3.97	3.67
Tyrosine	1.56	2.38	2.07
Tryptophan	1.07	1.94	1.33
Valine	4.92	5.15	5.37

[a]Janicki and Kowalczyk (1965).
[b]Ewart (1967).
[c]Tkachuk and Irvine (1969).

As with other cereals, rye storage proteins are characterized by having high contents of glutamic acid and proline. The aspartic acid content of rye protein is higher than that of wheat and barley but lower than that of oats and maize (Ewart, 1967). The amide content falls as the glutamic acid content decreases in the order wheat > rye, barley > oats > maize; the fraction of the total glutamic + aspartic present in the amide form follows a similar pattern. The sulfur-containing amino acids, although low, are not normally nutritionally limiting (Bixler *et al.,* 1968).

The distribution of side-chain groups in these cereals has been examined by Ewart (1967) and is summarized in Table IX.

Rye proteins closely resemble their barley counterparts in their distribution of hydrophobic, polar, and salt linkages. Wheat proteins, on the other hand, are characterized by having a high capacity for hydrogen bonding and dipole-dipole interactions (Thr + Ser + Tyr + amide) and the lowest capacity for ionic interactions or salt linkages. They also have a high ratio of ionic + polar to nonpolar groups. This makes the wheat storage protein complex the most hydrophilic of the five cereals examined and explains partially the solubility of some gliadin components in water and, to a smaller extent, in other aqueous solutions. There appears to be nothing about its amino acid composition which could account for rye having some dough-forming properties, while barley has none. Ewart (1967) suggests that in wheat, the weak generalized intermolecular associations, resulting from the combination of interactions of the side chains present, would give coherence to the proteins while still permitting viscous flow. However, there is a difference in the potential for -S-S-/-SH interchange, which may be correlated with the differences observed in rheological behavior (Redman and Ewart, 1967). In addition, the distributions of molecular weight among the proteins comprising the storage protein fraction in the different cereals and their capacity to undergo entanglement coupling (MacRitchie, 1972) are factors to which insufficient attention has been paid.

GENETIC AND ENVIRONMENTAL INFLUENCE ON PROTEIN AND AMINO ACID COMPOSITION

In a wide survey of Polish rye varieties, Trzebska-Jeske and Morkowska-Gluzinska (1963) found that while overall mean Kjeldahl nitrogen values were

TABLE IX

Side-chain groups in the flour proteins of rye, barley, corn, oats, and wheat (number of residues/10^5 g of recovered anhydro amino acids)[a]

	Rye	Barley	Corn	Oats	Wheat
Asp + Glu	275.9	267.7	203.3	239.8	312.7
Amide	182.7	184.5	111.5	138.4	248.7
Free COO⁻	93.2	83.2	91.8	101.4	64.0
Lys + His + Arg	69.0	67.8	82.3	93.3	53.1
Total ionic groups[b]	162.2	151.0	174.1	194.7	117.1
Excess of COO⁻	24.2	15.4	9.5	8.1	10.9
Polar (Thr + Ser + Tyr + amide)	287.5	287.8	226.2	255.0	354.7
Nonpolar groups	447.8	452.5	526.4	455.2	416.0
$\dfrac{\text{Ionic + polar}}{\text{Nonpolar}}$	1.00	0.97	0.76	0.99	1.13

[a]Source: Ewart (1967).
[b]At dough pH, ~ 6, the acidic and basic groups will be nearly all ionized.

not significantly different, within the same variety considerable variations in nitrogen content and 1000-kernel weight were observed. For example, the mean nitrogen content of the widely grown variety, Ludowe, was 1.44%, ranging between 1.30–1.67%. Tryptophan, methionine, and lysine contents (as percent of total protein) were inversely correlated with total protein, although their total amount (as percent of whole meal) increased with total protein content.

Boronoeva and Kazakov (1969) failed to find any substantial difference in amino acid composition between samples of winter and spring rye grown in different locations and seasons in the U.S.S.R. in spite of quite large differences in total protein content (10.8–15.3%) and 1000-kernel weight (16.1–33.9 g). This was confirmed by Somin (1970) and by Golenkov and Gilzin (1971) on rye samples grown in the U.S.S.R. under a variety of environmental conditions.

Primost *et al.* (1967) reported that the higher the nitrogen content of rye grown under varying conditions, the greater was the content of glutamic acid and proline (which are characteristic of the prolamin protein fraction) and the smaller was the content of lysine, arginine, and aspartic acid. Chlorcholine chloride (CCC) treatment, while affecting straw height, had no effect on amino acid composition.

Although not specifically reported for rye, it is likely that the effect of fertilizer treatment is similar to that on wheat. Thus, high applications of nitrogenous fertilizers raise protein levels in the mature wheat grain (Hutcheon and Paul, 1966; Swaminathan *et al.,* 1969; Dubetz, 1972; McNeal *et al.,* 1972). Where nitrogen supplies are limiting, an inverse relationship between grain yield and protein content has been noted by several workers (El Gindy *et al.,* 1957; Williams, 1966; Munck, 1964, 1972). Higher grain protein levels are generally associated with the presence of higher proportions of the storage protein fractions (Bell and Simmonds, 1963; Gandhi and Noltrawat, 1968; Abrol *et al.,* 1971); this is in agreement with the observation that higher protein levels in cereals are generally associated with a lower proportion of lysine (expressed as a percentage of the total protein) (McElroy *et al.,* 1949; Bendicenti *et al.,* 1957; Gunthardt and McGinnis, 1957; Lawrence *et al.,* 1958; McDermott and Pace, 1960; Sihlbom, 1962; Simmonds, 1962; Pereira, 1963; Sosulski *et al.,* 1963; Larsen and Nielsen, 1966; Deosthale *et al.,* 1969; Swaminathan *et al.,* 1969; Abrol *et al.,* 1971; Srivastava *et al.,* 1971; Johnson and Mattern, 1972). The explanation appears to be that higher protein levels are associated with a lower percentage of the water-soluble, lysine-rich groups of proteins, except in cases where specific mutations have occurred leading to a depression in the synthesis of the lysine-poor prolamin fraction and its replacement by glutelin and/or water-soluble protein fractions. Such mutations have been recognized in barley (HiProly) and maize (Opaque II and Floury), but not as yet in wheat or rye.

E. Fatty Acids

Total crude fat in rye, between 1.5 and 2.0%, resembles the amount in other cereals such as wheat, barley, and triticale but is much lower than that in oats. As shown in Table X, cereal lipids are characterized by having a high content of unsaturated fatty acids.

Rye differs from the other cereals listed in Table X by having a somewhat higher proportion of linolenic acid (18:3). This highly unsaturated fatty acid is

TABLE X

Total lipid and fatty acid composition of rye compared with other cereals and soybean[a]

	Crude Fat (% dry matter)	Fatty Acid Composition of Crude Fat (%)							
		Myristic 14:0	Palmitic 16:0	Stearic 18:0	Palmitoleic 16:1	Oleic 18:1	Linoleic 18:2	Linolenic 18:3	Eicosenoic 20:1
Rye	1.5 ± 0.2	...	16.5 ± 1.2	0.6 ± 0.1	...	15.6 ± 1.6	55.6 ± 2.3	10.4 ± 1.7	1.3 ± 0.3
Triticale	1.5 ± 0.2	...	18.7 ± 1.5	0.7 ± 0.2	...	13.4 ± 2.4	60.1 ± 1.5	6.3 ± 1.4	0.9 ± 0.1
Wheat	1.6 ± 0.2	0.1	17.8 ± 1.0	0.6 ± 0.1	0.8	14.5 ± 1.5	60.5 ± 1.4	5.7 ± 1.0	0.9 ± 0.2
Barley	1.7 ± 0.1	1.0	22.5 ± 1.3	1.0 ± 0.2	...	13.9 ± 1.1	54.8 ± 1.8	6.8 ± 0.4	1.0 ± 0.1
Oats	5.2 ± 0.4	...	16.3 ± 0.8	1.0 ± 0.2	...	40.2 ± 1.3	40.1 ± 0.9	1.7 ± 0.3	0.8 ± 0.1
Rice	2.2		17.6 (palmitic + stearic)			47.6	34.0	0.8	...
Soy	19.6		12.7 (palmitic + stearic)			27.7	53.7	5.9	...

[a]Source: Hulse and Laing (1974), Kent (1966), Markley (1950).

TABLE XI

Composition of rye grain, bran, and flour fractions (on a 15% moisture basis)[a]

		Rye Grain (g/100 g)	Bran (g/100 g)	Middlings	Low Grade Feed Flour
Dry matter		85	85	85	85
Ash		1.7	2.6	3.2	0.86
Crude fiber		2.5	2.8	5.7	0.95
Ether extract		1.2	1.9	2.9	1.1
Protein (N × 5.7)		9.9	14.2	14.7	8.9
Calcium		...	0.14	0.06	0.02
Phosphorus		...	1.22	0.59	0.32
Potassium		...	...	0.59	0.51
Manganese	mg/kg	...	11.2	41.5	...
Niacin	mg/kg	...	26.3	15.9	8.0
Pantothenic acid	mg/kg	...	15.9	21.8	9.8
Riboflavin	mg/kg	...	0.19	2.3	0.8
Thiamin	mg/kg	...	2.9	3.1	2.3

[a]Source: Recalculated from Feed Composition, Committee on Animal Nutrition (1964).

very sensitive to oxidation, which may lead to rancidity problems in the storage of rye flour. Lipids in cereals are derived from two principal sources: the storage lipids of the spherosomes in aleurone, scutellum, and germ tissue, and lipids associated with subcellular membrane structures in these tissues and the endosperm. The types of fatty acids associated with these two types of lipid differ. In addition, the nature of the fatty acids present in the storage lipids differs between cereal species. A considerable amount of further work is required to clarify the morphological distribution of lipids within the rye kernel.

F. Comparison of the Composition of Whole Rye Grain with that of Rye Bran and Flour

Data on the composition of rye grain and its bran and flour fractions have been compiled by the Canadian and American committees on animal nutrition and published jointly as tables of feed composition (Committee on Animal Nutrition, 1964). For comparison with similar English work presented below, their results have been recalculated to a 15% moisture basis and are summarized in Table XI.

Since there is no evidence that the bran and flour fractions were derived from the same rye sample used in the analysis of whole grain, it is not possible to draw conclusions from the differences in composition shown in Table XI. However, this may be done in the case of an extensive comparison of the chemical composition of wheat and rye grain and the flours derived from them at various extraction rates undertaken by McCance *et al.* (1945). From the previous discussion on the distribution of components between endosperm and bran, it will be clear that the differences in composition between whole grain and derived flour will depend strongly upon the percentage extraction employed in the milling process. For this reason, several extraction rates were used by McCance *et al.,* and the analyses for each were corrected to a 15% moisture basis. The rye used was of English origin but unknown variety, with protein content of 7.98%. This is considerably lower than the figures quoted in Table XI above, but is approximately in the middle of the range encountered with this cereal. Table XII summarizes the data for rye presented by McCance *et al.* (1945). Their very extensive data for wheat has also been summarized and is included in the table for comparison.

The trends in the case of both wheat and rye are similar. Progressive removal of the pericarp, aleurone, and outer endosperm in the bran fraction as the extraction rate decreases results in a reduction in the amounts of all components associated with the outer layers of the grain. The only component to increase in amount as a percentage of the flour is carbohydrate (mainly starch). Reference to the morphology of the grain, especially to Figure 9, makes it clear that the central endosperm cells, from which flours of low extraction are mainly derived, are particularly rich in starch. The changes in chemical composition observed are therefore those to be expected from the morphology of the grain.

The principal differences in behavior between wheat and rye were found to relate to the protein and fiber fractions. Sufficient protein was removed with the coarse bran in rye to bring its level in the 85% extraction flour well below that in the grain. The drop in protein content from whole rye meal to the finest flour was therefore considerably greater than in the case of wheat. Although rye contained less fiber, it was separated less cleanly, with the result that the fiber contents of

TABLE XII

Composition of whole rye and wheat and their milled products
(on a 15% moisture basis)[a]

Component:	English Rye				Manitoba Wheat				English Wheat				
% Extraction:	100	85	75	60	100	85	75	42	100	85	75	46	
Component:													
Total N	1.40	1.28	1.17	0.99	2.39	2.38	2.29	2.07	1.56	1.50	1.40	1.34	g/100 g
Protein (N × 5.7)	7.98	7.30	6.67	5.64	13.64	13.57	13.05	11.80	8.89	8.55	7.98	7.64	g/100 g
Fat	1.98	1.64	1.33	1.01	2.49	1.70	1.32	0.86	2.23	1.46	1.13	0.76	g/100 g
Carbohydrate (Starch)	69.0	73.0	75.0	78.0	63.0	67.2	69.5	71.2	66.8	72.0	74.2	75.8	g/100 g
Fiber	1.56	0.84	0.48	0.22	2.15	0.33	0.10	Trace	2.08	0.42	0.15	Trace	g/100 g
Thiamin	1.45	0.98	0.80	...	1.18	0.92	0.29	0.09	0.96	0.84	0.42	0.16	i.u./g
Riboflavin	2.90	2.00	1.40	0.85	1.70	1.00	0.70	0.50	1.70	1.20	0.60	0.50	μg/g
Ash	1.72	1.04	0.72	0.51	1.53	0.75	0.44	0.34	1.52	0.70	0.46	0.37	g/100 g
Potassium	412	203	172	140	312	146	87	71	361	179	118	99	mg/100 g
Calcium	31.5	26.1	19.5	15.3	27.6	18.5	13.1	11.1	35.5	24.5	19.2	15.2	mg/100 g
Magnesium	92	45	26	16	141.0	61.8	30.4	21.5	106.0	35.0	16.8	8.7	mg/100 g
Iron	2.70	1.97	1.72	1.32	3.81	...	...	...	3.05	2.22	1.35	0.95	mg/100 g
Total phosphorus	359	193	129	78	350	188	109	82	340	153	93	68	mg/100 g
Phytate phosphorus	258	104	57	24	242	96	37	14	233	73	30	10	mg/100 g

[a]Source: McCance et al. (1945).

TABLE XIII
Amino acid composition of rye grain, rye flour, and bran[a]

	Rye Grain		Sieved Flour		Break Flour		Bran	
	Total Protein %	Product %	Total Protein %	Product %	Total Protein %	Product %	Total Protein %	Product %
Alanine	5.13	0.55	4.22	0.41	4.78	0.51	5.36	0.88
Ammonia	2.92	0.30	2.74	0.28	2.16	0.22	2.19	0.37
Arginine	5.62	0.60	4.57	0.44	4.94	0.53	6.31	1.02
Aspartic acid	7.16	0.77	6.12	0.58	7.16	0.80	7.47	1.22
Cystine/Cysteine	1.19	0.13	1.20	0.13	0.91	0.08	1.90	0.32
Glycine	4.79	0.52	3.73	0.36	4.84	0.52	5.44	0.89
Glutamic acid	29.91	3.25	34.46	3.22	34.94	4.03	27.93	4.53
Histidine	2.09	0.22	1.90	0.19	1.82	0.18	2.19	0.35
Isoleucine	3.88	0.42	3.10	0.30	4.09	0.44	3.69	0.61
Leucine	7.00	0.75	5.85	0.56	7.33	0.81	6.76	1.09
Lysine	4.23	0.46	3.30	0.32	3.46	0.37	4.06	0.67
Methionine	0.65	0.07	0.21	0.02	1.08	0.10	0.44	0.08
Phenylalanine	5.25	0.56	4.99	0.48	5.57	0.62	4.56	0.73
Proline	5.20	0.55	5.90	0.56	5.96	0.65	4.93	0.79
Serine	4.54	0.48	4.56	0.44	5.05	0.55	4.53	0.73
Threonine	3.11	0.34	2.54	0.25	3.29	0.35	3.34	0.57
Tyrosine	2.86	0.31	2.67	0.26	2.90	0.31	2.66	0.43
Valine	5.56	0.60	4.92	0.48	5.45	0.60	5.32	0.86

[a]Source: Rukosuev and Silant'eva (1972) via Hulse and Laing (1974).

rye flours generally were higher than those of their wheat counterparts.

Compared with the flour, rye bran contains higher amounts of lysine, histidine, arginine, glycine, alanine, and the sulfur-containing amino acids cystine and methionine (Table XIII). In addition, it has lower amounts of glutamic acid, proline, and phenylalanine. The storage protein of the aleurone layer therefore differs in nature and composition from that of the endosperm.

IV. PROTEIN COMPOSITION OF THE MATURE RYE KERNEL

A. Protein Distribution in the Kernel

PERICARP

As discussed in earlier sections of this chapter, the distribution of protein in the rye kernel is related to its function in the development and germination of the grain. Thus, the protein content of the outer pericarp at maturity is low, reflecting the degeneration of these layers during grain development and the hydrolysis and absorption of the cytoplasmic protein components in the cells of the outer pericarp, outer and inner integument, and nucellar epidermis. At maturity, less than 4% of the total grain protein is present in the pericarp. This protein is highly insoluble and is probably complexed with tannin components.

ALEURONE AND ENDOSPERM

The aleurone and endosperm both serve as storage organs in the grain. However, they serve different functions on germination. The reserves contained in the aleurone cells comprise the spherosomes, a source of lipid and energy, and the phytin granules. The latter represent a rich source of phosphorus for the production of adenosine triphosphate and other high energy phosphates, together with soluble, readily metabolized storage protein. Other proteins in the aleurone cells are of the cytoplasmic type, occurring in the cell cytoplasm and concerned with its metabolism both during kernel development and upon germination. The function of the aleurone cells is to provide, on germination, an instant source of energy and raw materials (amino acids) for the synthesis of enzymes directed towards the major food reserves of the grain, starch and protein, which are stored in the endosperm.

The data in Tables XI and XII indicate that the aleurone cells are comparatively rich in protein since this layer contributes almost exclusively to the high protein content of the bran fraction obtained on milling. Reliable data for the protein content of aleurone cells isolated by hand dissection of rye grain are lacking; however, from Table XI it may be calculated that they contain at least 16.7% protein dry weight and probably over 20%.

The endosperm, on the other hand, contains high proportions of starch as a storage material, and its protein content is considerably lower than that of the aleurone layer. From Tables XI and XII, it may be seen that protein levels of 6–9% are representative of this portion of the grain.

B. Extraction, Fractionation, and Properties of Rye Endosperm Proteins

While a certain amount of work has been carried out on the extraction, fractionation, and immunoelectrophoretic properties of wheat aleurone tissue (summarized by Simmonds and Orth, 1973), no comparable studies have been carried out on rye aleurone proteins. This section will therefore be concerned exclusively with work on the cytoplasmic and storage proteins of rye endosperm cells.

The classical procedures developed by Osborne (1907) for the fractionation of cereal proteins have been applied to rye, durum wheat, hexaploid wheat, and triticale by Yong and Unrau (1966) and Chen and Bushuk (1970a,b,c). These procedures suffer from two main disadvantages: different protein classes overlap somewhat in their solubility characteristics, and no fractionation scheme has yet been devised which completely solubilizes all protein present in the residue, which may sometimes contain up to 35% insoluble protein. Within these limitations, however, rye endosperm proteins, in common with those of other cereals, may be divided according to their solubility characteristics into two principal groups:

(i) Cytoplasmic, or Metabolically Active, Proteins. These are soluble either in water (albumins) or dilute salt solutions (globulins) and comprise those enzymes and related proteins which were largely responsible for the metabolic activity of the developing cell or for its utilization as a source of food and energy upon germination. They are located in the cytoplasm, and in wheat at least, their concentration around the starch granules has been claimed to play an important role in determining grain hardness (Barlow *et al.*, 1973; Simmonds *et al.*, 1973).

(ii) Storage Proteins. These are generally insoluble in water and dilute salt solutions. They have been differentiated by Osborne into prolamins, soluble in 70% ethanol, and glutelins, soluble in dilute acids or alkalis. Unfortunately, progressive extraction of flour with water, salt solutions (*e.g.*, 0.5M sodium chloride), 70% ethanol, and 0.05M acetic acid results in between 20 and 35% of the total protein remaining unextracted in the final residue. The introduction of other solvents, *e.g.*, acetic acid-urea-cetyltrimethylammonium bromide (AUC) (Meredith and Wren, 1966), has improved, but not eliminated, this problem.

Figure 10 shows the distribution of protein into the four solubility classes listed above and the residue material for the four cereal species investigated by Chen and Bushuk (1970a,b,c).

The rye sample contains much higher amounts of the water- and salt-soluble proteins and significantly lower amounts of the alcohol-soluble, acetic acid-soluble, and residue proteins than the hard red spring (HRS) wheat sample. This distribution of the major protein groups in rye, as compared with bread wheats, is reflected in its overall amino acid composition. This aspect was studied in greater detail by Yong and Unrau (1966) who analyzed the solubility classes derived from each of the four cereal species.

Glutamic acid and proline were the predominant residues in all fractions. However, compared with the alcohol- and acid-soluble (or storage) protein group, the water- and salt-soluble fractions contained somewhat less of these amino acids. Lysine was particularly low in the alcohol-soluble storage protein fractions of all the cereals examined, ranging between 0.34–0.79 mol %. In the acid-soluble group, the lysine content was higher (1.37–2.03 mol %) while in the salt- and water-soluble groups it was higher again (2.03–2.64 and 2.54–3.68 mol %, respectively). In all cases rye had the lowest lysine content in each protein group. However, this was more than compensated for by the increased amount of water-soluble protein present in this cereal so that the overall lysine content of rye flour is higher than that of the other cereals.

Proteins in the groups fractionated according to solubility were further resolved by gel filtration and acrylamide gel electrophoresis (Chen and Bushuk, 1970b,c). Each group was shown to consist of several protein species, and it is likely that if two-dimensional electrofocusing-electrophoresis techniques had been applied (Wrigley, 1970), resolution of additional components would have been achieved.

Fractionation on the basis of molecular size was chosen by Chen and Bushuk as the more reliable guide to distinguishing between the four protein classes. They suggest the following size ranges (Table XIV) as being representative of the proteins in each solubility class. These correspond closely to those suggested for wheat proteins by Meredith and Wren (1966). However, it must be stressed that extractions and chromatography were carried out using dissociating solvents so that the size ranges obtained are likely to be minimum values.

On gel filtration, each of the solubility fractions obtained from all four cereals were shown to be contaminated by significant amounts of at least one other protein class. In particular, the albumin contents in Figure 10 are overestimates, since in all cases this fraction contained a considerable proportion (28–45%) of high-molecular-weight material which emerged at the hold-up volume of the column and was probably derived from the prolamin and glutelin fractions. Nevertheless, the albumin content of rye was still higher than that of the other three cereals. Further examination of the protein classes, after purification by gel filtration, showed that each was heterogeneous as judged by disc gel

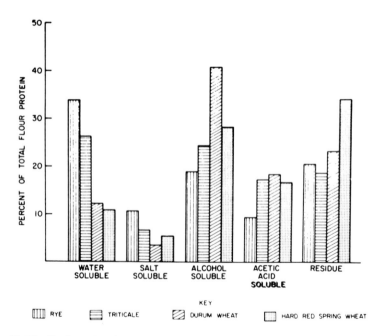

Figure 10. Distribution of endosperm proteins according to solubility characteristics in rye, triticale, durum, and hard red spring (HRS) wheat (Chen and Bushuk, 1970a).

electrophoresis at pH 3.8. Generally about five electrophoretic components were resolved in each fraction.

Fractionation of rye proteins can also be obtained by extraction with $2M$ urea which removes the water-soluble proteins and prolamins and leaves the glutelins in the residue (Shepherd, 1968, 1973; Wrigley, personal communication). Starch gel electrophoresis at pH 3, of the $2M$ urea extract, can then be used to separate the rye prolamins as a group of components with much lower mobilities than the water-soluble proteins (Figure 11). Further extraction with $6M$ urea dissolves much of the rye glutelins from the residue. Being of high molecular weight, these do not enter the starch gel and are retained at the origin. Slightly different electrophoretic patterns for the prolamin proteins are obtained from different rye genotypes (Ellis, 1971). In addition, the patterns given by rye prolamins differ distinctly from those of wheat gliadins, as shown in Figure 11.

The electrophoretic patterns of Figure 11 can be usefully employed in distinguishing between rye and wheat flours (and even baked goods) and in determining the approximate composition of a wheat-rye grist. Even so, a basic similarity in the prolamins of wheat and rye is indicated by peptide mapping (Ewart, 1966a) and immunological comparisons (Ewart, 1966b).

Gel electrofocusing of rye prolamins shows that their isoelectric points lie mainly between about 5 and 9 (Wrigley and Shepherd, 1973). Combination of gel electrophoresis with electrofocusing reveals that many of the zones separated by the former method contain more than one protein. Thus, the two-dimensional combination of the methods (Figure 12) increases the estimate of eight bands in King II rye obtained by electrophoresis to over 20 components. The combination of electrofocusing and electrophoresis also reveals a greater heterogeneity among the water-soluble proteins than does electrophoresis alone.

TABLE XIV
Molecular weight ranges (mol wt) of proteins extracted
from rye, triticale, durum, and hard red spring (HRS) wheats

Solubility Class	Chen and Bushuk (1970b) mol wt	Meredith and Wren (1966) mol wt
Albumins and globulins	10,000–30,000	10,000–25,000
Prolamins	50,000–90,000	25,000–100,000
Glutelins	> 150,000	> 100,000

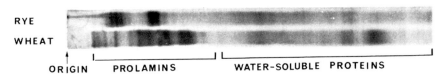

RYE

WHEAT

ORIGIN   PROLAMINS   WATER-SOLUBLE PROTEINS

Figure 11. Starch gel electrophoresis (pH 3) of proteins extracted from rye flour and wheat flour with $2M$ urea, according to the method of Wrigley and Shepherd (1975).

GENETIC CONTROL OF THE SYNTHESIS OF RYE ENDOSPERM PROTEINS

Shepherd (1968, 1973) examined the starch gel electrophoretic behavior of sodium pyrophosphate and 2*M* urea extracts of King II rye (2n = 14), Holdfast wheat (2n = 42), the Holdfast + King II amphiploid (2n = 56), and seven separate wheat-rye addition lines (2n = 44). It was found that the rye prolamin proteins were also extracted by sodium pyrophosphate (0.01*M*, pH 7.4) so that these extracts contained a full complement of rye albumin and globulin bands as well as seven distinct prolamin bands (Shepherd and Jennings, 1971). By comparing the electrophoretic patterns of the seven addition lines with that of the parent rye

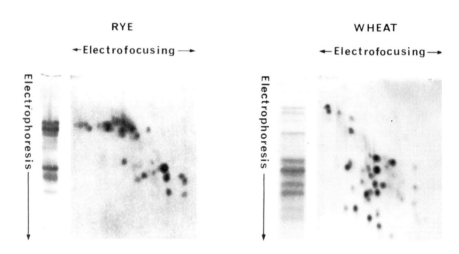

RYE   WHEAT

←Electrofocusing→   ←Electrofocusing→

Electrophoresis

Figure 12. Combined gel electrofocusing (pH 5, left, to 9) and gel electrophoresis of prolamins from rye (cv. King II) and wheat (cv. Bungulla). The pattern for starch gel electrophoresis alone appears on the left in each case. The patterns for rye are reproduced from Wrigley and Shepherd (1973).

TABLE XV
Characteristics of α-amylases isolated from germinating rye, durum wheat, and triticale[a]

Characteristic	Rye (cv. Prolific)	Triticale (cv. 6A190)	Durum Wheat (cv. Stewart)
Electrophoretic mobility ($cm^2\ sec^{-1}\ V^{-1} \times 10^5$)	2.05–2.08	2.39–2.43	2.31–2.36
pH optimum	4.9–5.0	4.7–4.8	4.5–4.6
pH stability range	4.0–9.1	3.9–7.9	4.0–8.4
Temperature optimum	50°–52°C	54°–56°C	54°–56°C
Energy of activation: (a) 10°–20°C	8.2 kcal/mol	10.3	13.5
(b) 30°–40°C	3.1	5.4	8.1
Energy of heat inactivation: 70°–80°C	13.7	17.0	40.6
Km (mg starch/ml)	2.78	6.25	7.15
Vm (mg maltose-ml/3 min)	1.10	1.00	0.62

[a]Source: Lee and Unrau (1969).

and wheat and the amphiploid between them, Shepherd concluded that in rye, only chromosome V carried genes controlling the synthesis of prolamins. Furthermore, since the pattern of the chromosome V addition lines was identical with that of the short-arm ditelocentric addition line and the amphiploid, it would appear that all of the electrophoretically slow-moving proteins in rye are controlled by genes located on the short arm of chromosome V. Similar studies on addition lines involving Imperial rye and Chinese Spring wheat have indicated that chromosome E of Imperial rye was solely responsible for prolamin synthesis in that variety. Subsequently, chromosomes designated V and E were shown to be homologous and are now designated 1R, since they are also homoeologous with the wheat chromosomes of group 1. In this respect, the situation in rye differs from that in wheat, where two chromosome groups, numbers 1 and 6, have been shown to carry the genes controlling the synthesis of the gliadin proteins (Wrigley and Shepherd, 1973). On the basis of these results, it has been suggested that the genes controlling prolamin synthesis in the original diploid ancestor of the Triticinae were located on only one of its seven chromosomes. However, subsequent to the differentiation of rye, an ancestral form of wheat acquired the two-chromosomal control of prolamin synthesis.

C. Enzymatic Activities in the Rye Kernel

All cereal grains at maturity show low, but significant, amounts of enzymatic activity, directed particularly towards the principal storage components, starch and protein. Rye is no exception, and particular attention has been focused on changes in α-amylase activity during kernel development, at maturity, and upon germination. However, it must be recognized that during these phases of growth, the active tissues of the grain, embryo, aleurone, endosperm, and pericarp will contain the complete complement of enzymes necessary for tissue growth or autolysis. During this cycle, the mature grain represents a dormant phase in which enzymatic activity is at a minimum. Because they represent the major components at maturity, the endosperm and aleurone cells tend to dominate the picture at this stage. During grain development, however, particular tissues may in turn contain major proportions of specific enzymes whose activities may decline at maturity.

α-AMYLASE

Purified preparations of this enzyme were obtained from germinating rye, wheat, and triticale by Lee and Unrau (1969), who also compared their amino acid compositions, gel electrophoretic mobilities, and enzymatic activities. The enzymes were purified by ammonium sulfate and acetone fractionation, gel filtration, and refiltration of active fractions, followed by preparative electrophoresis of active fractions on polyacrylamide gel. Ultracentrifugation indicated the presence of a single component of identical sedimentation coefficient for the enzymes isolated from the three cereal species. A mixture of the three enzyme preparations gave one elution peak on gel filtration, but it could be partially resolved into three bands by gel electrophoresis. Significant differences were also observed in amino acid composition, temperature and pH responses, and the Michaelis-Menten constants K_m and V_m. Several of these properties are summarized in Table XV.

The levels of α-amylase activity were generally very low in ungerminated grains but increased rapidly and logarithmically on germination, the rate of increase in the case of rye being particularly rapid (Lee and Unrau, 1969).

Changes in α-amylase activity in the developing kernel have been studied by Klassen *et al.* (1971) and Dedio *et al.* (1974). In the four cereals examined, α-amylase activity was found to vary widely during development but to be at a minimum at maturity. However, the activities in rye and triticale were both considerably higher than those in tetraploid and hexaploid wheats, confirming the earlier observations by Müntzing (1963).

Dissection of the grain into pericarp, aleurone, and endosperm tissue reveals that at least three different α-amylase activities are involved at various periods of development (Dedio *et al.*, 1974). Figure 5b shows the appearance of starch granules in the thin-walled parenchyma cells of the pericarp at about 4–5 days after anthesis. These become eroded and disappear during the next 2–3 weeks, during which time α-amylase activity in the pericarp is at a maximum, as shown in Figure 13.

Subsequently, the activity in the pericarp diminishes, and at maturity it is quite low. Activity in the aleurone layer fluctuates rather less during grain

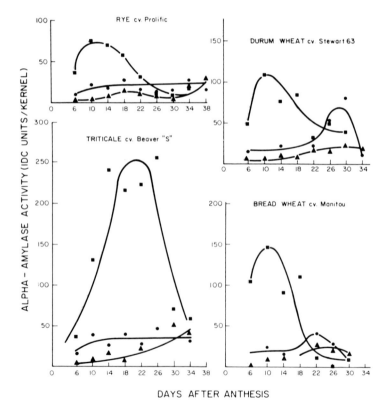

Figure 13. Distribution of α-amylase activity in pericarp (■), aleurone (●), and endosperm (▲) of rye, triticale, durum wheat, and bread wheat kernels during development. (Constructed from data of Dedio *et al.*, 1974.)

development but also tends to diminish at or towards maturity, provided moist conditions are not encountered which lead to premature germination. In the case of some triticale lines, however, premature release of α-amylase from about 21 days after anthesis onwards is associated with degeneration of the aleurone layer and necrotic areas in the endosperm (Simmonds, 1974). In rye, activity in the endosperm increases slightly from 18 days after anthesis and at maturity represents the major contribution to the final activity of the grain. In tetraploid and hexaploid wheats, α-amylase activity in the endosperm generally falls off as maturity is reached. It should be stressed that the total levels encountered at maturity in the absence of pregermination are extremely low compared with those liberated by the aleurone cells upon germination.

β-AMYLASE

Low concentrations of this enzyme have been reported to be present in most cereals at maturity. Little work has been carried out on the rye enzyme.

PROTEASE

Comparatively little work has been reported on the proteolytic enzymes present in rye. They appear to be similar to those in wheat, being present in higher concentrations in the bran layers and being extractable by dilute salt solutions of pH 4.5–7. Acetate buffers (0.1–0.2M; pH 4.45–5.0) are frequently used for this purpose. Rye protease was found by Breyer and Hertel (1974) to have optimum activity at pH 8.0 and 50°C, compared with the wheat enzyme, which was most active at pH 7.0 and 33°C. These authors were using principally the synthetic substrate, α-benzoyl-DL-arginine-4-nitro-anilide (BAPA). On the other hand, Madl and Tsen (1973a), comparing the proteases of rye, triticale, and wheat, found them to have essentially the same pH optimum at 4.45 when tested against hemoglobin as a substrate.

It is seen in Table XVI that the protease activities of the rye samples are comparable with those of the triticales and generally higher than those of the wheats. Madl and Tsen noted that no direct correlation between the protein content and proteolytic activity was observed. In fact, there is probably no reason why such a correlation should exist.

ESTERASE

Esterase activity in developing rye endosperm has been used by Barber *et al.* (1968) as a marker for homology between wheat and rye chromosomes. The enzyme was found in both leaf and developing grain tissue. The enzyme from the developing endosperm of rye gave a single band while that from wheat gave a group of three anodically fast-moving bands. The octoploid triticale line examined by these authors was found to contain five fast-moving bands, four of them corresponding to the parental wheat and rye components and an additional one appearing between band 1 (from rye) and band 3 (from wheat). Further genetic analysis indicated that the genes coding for the single esterase band in rye were carried on chromosome G (= 3R).

β-GLUCOSIDASES

The β-glucosidase and β-galactosidase activities of various cereals at maturity have been compared by Gelman (1969). The enzymes were extracted from wholemeal preparations, using 0.2N acetic acid-sodium acetate, pH 5.0

buffer, and were tested against the o-nitrophenol derivatives of glucose and galactose as substrates. Inhibition and heat inactivation studies showed the enzymes derived from the different cereals to be essentially similar (Table XVII). Both β-glucosidase and β-galactosidase activities were found to increase on germination.

V. STARCH COMPOSITION OF THE MATURE RYE KERNEL

The starch content of rye flour is not greatly different from that of flours derived from tetraploid and hexaploid wheats. However, the individual granules differ in size distribution from their wheat counterparts, and the amylograph viscosity of rye flour is much lower due to the presence of significant amounts of α-amylase activity in the kernels at maturity (Klassen and Hill, 1971, and

TABLE XVI

Relative proteolytic activities of flours and brans from wheat, rye, and triticale[a]

	Whole Grain		Flour		Bran	
	% Protein (d.b.)[b]	% Flour Yield	% Protein (d.b.)[b]	Activity[c]	% Protein (d.b.)[b]	Activity[c]
Eagle[d] (wheat)	12.7	70	11.8	1.74	14.3	4.31
Centurk[d] (wheat)	12.7	73	11.4	1.75	15.2	4.87
Scout-1[e] (wheat)	13.6	72	13.1	1.73	16.7	5.37
Scout-M[f] (wheat)	14.6	72	13.9	1.65	17.7	4.80
Minn. 11[g] (rye)	10.0	51	7.1	2.61	13.9	6.48
Balbo[d] (rye)	12.6	45	8.1	2.07	15.7	5.77
Balbo[f] (rye)	16.3	48	12.7	2.08	19.5	5.75
Triticale-1[f]	13.7	64	12.6	1.98	18.1	6.60
Triticale-F[e]	15.1	61	14.1	2.01	18.6	6.15
Triticale-385[e]	15.4	62	13.6	2.00	20.0	5.70
Triticale-298[f]	17.5	58	15.5	1.93	21.1	5.96
Armadillo–Triticale[h]	16.2	67	13.9	2.68	21.3	6.35
Bronco–Triticale[h]	19.9	69	17.6	2.53	23.3	6.45

[a]Source: Madl and Tsen (1973a).
[b]Protein determined as (% N × 5.7).
[c]Expressed as mg protein per ml of supernatant.
[d]Harvested from Manhattan, Kansas, experimental plots in 1972.
[e]Harvested from Garden City, Kansas, experimental plots in 1971.
[f]Harvested from Manhattan, Kansas, experimental plots in 1971.
[g]Harvested from Columbia, Mo., experimental plots in 1972.
[h]Harvested from Mexican experimental plots in 1971.

TABLE XVII

Comparison of β-glucosidase and β-galactosidase activity in various cereals at maturity[a]

	β-Glucosidase[b]	β-Galactosidase	Ratio
Rye	25,000	12,000	2.1
Barley	27,000	7,000	3.8
Oats	6,000	10,000	0.6
Wheat	18,000	8,000	2.3

[a]Source: Gelman (1969).
[b]μg o-nitrophenol liberated per g dry cereal.

previous section). The properties of the individual isolated starches have been investigated by Berry *et al.* (1971) and by Klassen and Hill (1971). Some of their results are summarized in Table XVIII which compares rye starch with those isolated from hexaploid triticale and bread and durum wheats. The exact varieties on which the measurements were carried out were not stated by Berry *et*

TABLE XVIII

Properties of rye starch compared with starches prepared from hexaploid wheat (hard red spring), tetraploid wheat, and triticale

Sample	Rye	Triticale	HRS Wheat	Durum Wheat
Property:				
Starch recovery (%)	42	53	55	54
Nitrogen (%)	0.04	0.04	0.04	0.05
Ash (%)	0.37	0.39	0.38	0.49
Fat (%)	0.98	0.78	0.75	0.70
Phosphorus (%)	0.025	0.042	0.047	0.047
Absolute density at 30°C, g per cc[a]	1.4209	1.4465	1.4832	1.4460
Water-binding capacity (%)[b]	86.5	...	85.0	85.5
Intrinsic viscosity[b]	1.96	2.13	1.88	2.10
Amylose (%)[b,c]	24.0	23.0	24.5	27.0
Amylose (%)[c,d]	30.1	23.7	28.9	30.1
Sedimentation coeff. ($S_{20,w.}$)[d]	3.49	3.63	3.42	3.17

[a] As is basis.
[b] Dry basis.
[c] Berry *et al.* (1971).
[d] Klassen and Hill (1971).

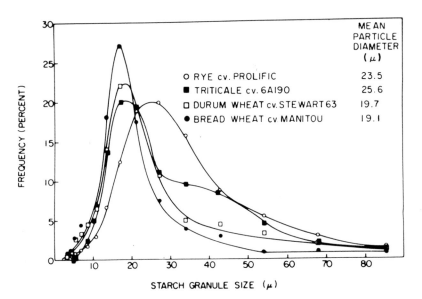

Figure 14. Mean particle diameter (M.P.D.) and size distribution of starch of rye cv. Prolific, triticale cv. 6A190, durum wheat cv. Stewart 63, and bread wheat cv. Manitou. (From data of Klassen and Hill, 1971.)

al. (1971); however, the figures in Table XVIII may be taken as representative.

Not only is the mean particle diameter of rye starch greater than those of the other cereals examined, but its size distribution covers a wider range (Berry *et al.,* 1971; Klassen and Hill, 1971). The results obtained by Klassen and Hill are presented in Figure 14.

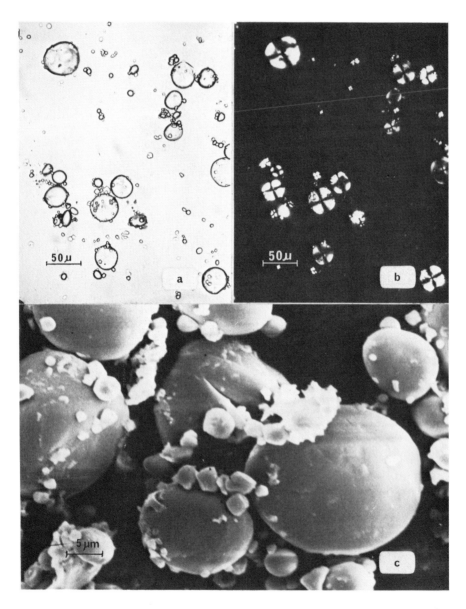

Figure 15. Rye starch granules (a) under bright field illumination, (b) under polarized light ordinary microscope, and (c) under the scanning electron microscope.

The lenticular shape of the A-type granules is readily apparent under the light microscope (both bright field and under polarized light) and under the scanning electron microscope (Figure 15). Smaller, spherical B-type granules are not as common in rye as in wheat and triticale. This confirms the particle size distribution curves shown in Figure 14. The development of rye, wheat, and triticale starch granules from anthesis to maturity has been studied by Dronzek *et al.* (1974) and compared with earlier studies on wheat and barley starch by Sandstedt (1946), May and Buttrose (1959), Buttrose (1960, 1963), and Evers (1971).

The developmental sequence observed was similar, with differentiation of the larger A-type granules first being noted at 5–7 days after anthesis. These granules appeared to arise from the smaller, spherical type. The latter served as nuclei on which more starch was deposited to build up preferentially in the equatorial plane. At about 7 days after anthesis, some granules appeared to be oblong or bean-shaped. Dronzek *et al.* (1974) suggest that the outermost portions of these ring structures grow towards each other in the equatorial plane and eventually fuse. However, it is not easy to find evidence in the mature granule for any fusion rings or zones which might correspond to this mode of granule development.

Undamaged rye starch granules, in common with other starches, exhibit birefringence under polarized light. This property disappears when the granules are damaged or gelatinized by heat. The birefringence endpoint temperature (BEPT) has been used as a measure of the completion of gelatinization. The BEPT and gelatinization temperature range of rye starch are similar to those of wheat and triticale (Table XIX).

Rye and triticale flours give very low peak amylograph viscosities compared with those of hexaploid and tetraploid wheat flours. This was shown by Klassen and Hill (1971) to result from high α-amylase activity in the mature rye and triticale kernels. By conducting the test in the presence of 200 μM AgNO$_3$, which inhibited the action of the enzyme, values for rye and triticale were obtained which approached those of wheat flour. Amylograms of the isolated starch granules also confirmed that the effects were primarily enzymatic in origin since the peak viscosities of rye starch approached, and of triticale starch exceeded, those of the bread and durum wheats examined by Berry *et al.* (1971).

Prolific rye starch was reported by Klassen and Hill (1971) to contain a higher linear amylose fraction than either triticale or hexaploid wheat.

TABLE XIX
Birefringence endpoint temperatures (BEPT) and gelatinization
temperature ranges of rye, wheat, and triticale starches[a]

Sample	BEPT (°C)	Gelatinization Temperature (°C)
Rye cv. Prolific	59.6	9.6
Hard red spring wheat cv. Manitou	62.9	8.6
Durum wheat cv. Stewart 63	56.1	10.2
Triticale cv. 6A190	61.5	6.3

[a]Source: Klassen and Hill (1971).

This was not confirmed by the results of Berry *et al.* (1971), and both sets of figures have been included in Table XVIII for comparison. The figures obtained by Klassen and Hill for the sedimentation coefficients in the ultracentrifuge of amylose fractions obtained from the four starches are also included in Table XVIII.

Comprehensive studies on the chemical structures of the starches of rye, triticale, bread wheat, and durum wheat were also carried out by Hew and Unrau (1970) and by Berry *et al.* (1971). These studies confirmed that as in its physical properties, rye starch is generally not greatly different from starches isolated from other related cereal species (Table XX).

The kinetics of hydrolysis and amount of α- and β-limit dextrins obtained by α- and β-amylolysis of rye starch were essentially the same as those obtained from wheat and triticale starches (Hew and Unrau, 1970), again confirming the similarity of structure between these starches.

VI. CONCLUSION

In morphology, chemical composition, and physical properties, the rye kernel shows an overall similarity to other cereals. This similarity extends to the developmental morphology of the grain, the types of proteins present in its endosperm, and the chemical structure and physical properties of the starch granules it contains. The question naturally arises: What is it that sets wheat, rye, and triticale apart in conferring upon their flours the breadmaking properties which have become so important technologically?

Grain hardness plays a major role in determining milling behavior, and as a consequence, the extent of starch damage, water absorption, and gassing power of the derived flour. Nevertheless, it is generally considered that the ability to form a cohesive dough having breadmaking characteristics is a property of the protein fraction. Both qualitative and quantitative differences exist between cereals in the distribution of the different protein classes present in this fraction. However, no one property has yet been demonstrated to distinguish wheat, rye, and triticale from the nondough-forming cereals. It would appear from the work of many investigators that the answers, when they emerge, will be found in an examination of three main areas:

TABLE XX

Chemical structure, molecular weight, and intrinsic viscosity of amylose and amylopectin fractions isolated from rye, triticale, and wheat starches[a]

Sample	Fraction	Branching %	Glucose Units in a Segment	Molecular Weight	Intrinsic Viscosity
Rye	Amylose	...	...	218,420	2.60
Triticale	Amylose	...	...	261,713	2.86
Hard red spring	Amylose	...	...	284,840	2.78
Durum wheat	Amylose	...	...	272,829	2.87
Rye	Amylopectin	4.8	21	...	1.33
Triticale	Amylopectin	4.5	22	...	1.78
Hard red spring	Amylopectin	4.4	23	...	1.83
Durum wheat	Amylopectin	4.8	21	...	1.96

[a]Source: Berry *et al.* (1971).

(a) The detailed molecular weight distribution of the proteins contributing dough-forming properties to the flour. This implies that not only must a wide range of molecular weights be covered by the proteins present, but also that the quantitative distribution of protein species throughout that range is important, if not critical.

(b) The numbers and distribution of polar, charged, and nonpolar side chains in the protein complex. This distribution will determine both the solubility properties as well as the proportion and extent of hydrogen bonding and ionic and hydrophobic interactions.

(c) The presence of a restricted number of rheologically important disulfide and sulfhydryl groups. The presence of these groups and their relative proportions may well be critical, since in spite of the low number present, they exert a major influence over the distribution of molecular weight in the final protein complex.

A fascinating and challenging prospect lies ahead in the further investigation of these three areas and in the close comparison of the breadmaking cereals with those lacking these properties. The pentosans of rye and their interaction with the proteins during dough formation deserve particular attention.

VII. ACKNOWLEDGMENTS

It is a pleasure to acknowledge the assistance of Miss A. E. Bonnefin in the preparation of the figures. Dr. C. W. Wrigley performed the electrofocusing and electrophoretic separations reported in Figures 11 and 12 and assisted in many ways in discussions of the text.

LITERATURE CITED

ABROL, Y. P., UPRETY, D. C., AHUJA, V. P., and NAIK, M. S. 1971. Soil fertilizer levels and protein quality of wheat grains. Aust. J. Agr. Res. 22: 195-200.

BARBER, H. W., DRISCOLL, C. J., and VICKERY, R. S. 1968. Enzymic markers for wheat and rye chromosomes. Proc. 3rd Int. Wheat Genet. Symp., Canberra, 1968, ed. by K. W. Finlay and K. W. Shepherd; p. 116. Butterworth: Australia.

BARLOW, K. K., BUTTROSE, M. S., SIMMONDS, D. H., and VESK, M. 1973. The nature of the starch-protein interface in wheat endosperm. Cereal Chem. 50: 443-454.

BELL, P. M., and SIMMONDS, D. H. 1963. The protein composition of different flours and its relationship to nitrogen content and baking performance. Cereal Chem. 40: 121-128.

BENDICENTI, A., BOGILIOLO, M., MONTENERO, P., and SPADONI, M. A. 1957. Amino acid content of some varieties of Italian wheat of varying protein contents (in Italian). Quad. Nutr. 17: 149-158.

BERRY, C. P., D'APPOLONIA, B. L., and

GILLES, K. A. 1971. The characterization of triticale starch and its comparison with starches of rye, durum and HRS wheat. Cereal Chem. 48: 415-427.

BIXLER, E., SCHAIBLE, P. J., and BANDEMER, S. 1968. Preliminary studies on the nutritive value of triticale as chicken feed. Quart. Bull. Mich. Agr. Exp. Sta. 50: 276-280.

BONNETT, O. T. 1966. Inflorescences of maize, wheat, rye, barley and oats: Their initiation and development. Ill. Agr. Exp. Sta. Bull. 721: 49-58.

BORONOEVA, G. S., and KAZAKOV, E. D. 1969. Amino acid composition of spring rye grain (in Russian and English). Prikl. Biokhim. Mikrobiol. 5: 314-317.

BRADBURY, D., MacMASTERS, M. M., and CULL, I. M. 1956. Structure of the mature wheat kernel. II. Microscopic structure of pericarp, seed coat, and other coverings of the endosperm and germ of hard red winter wheat. Cereal Chem. 33: 342-360.

BRENCHLEY, W. E. 1909. On the strength and development of the grain of wheat *(Triticum vulgare).* Ann. Bot. 23: 117-139.

BREYER, D., and HERTEL, W. 1974. The determination of proteolytic activities by synthetic substrates in wheat and rye and their milling products. Getreide, Mehl Brot. 28: 13-16.

BUTTROSE, M. S. 1960. Submicroscopic development and structure of starch granules in cereal endosperms. J. Ultrastruct. Res. 4: 231-257.

BUTTROSE, M. S. 1963. Ultrastructure of the developing wheat endosperm. Aust. J. Biol. Sci. 16: 305-317.

CAMPBELL, W. P., LEE, J. W., and SIMMONDS, D. H. 1974. Protein synthesis in the developing wheat grain. Proc. 24th Annu. Conf. Roy. Aust. Chem. Inst., Cereal Chem. Div. p. 6.

CARTTER, J. L., and HOPPER, J. H. 1942. The influence of variety, environment, and fertility level on the chemical composition of soybean feed. U.S. Dep. Agr. Tech. Bull. 787, p. 1-66.

CHEN, C. H., and BUSHUK, W. 1970a. Nature of proteins in triticale and its parental species. I. Solubility characteristics and amino acid composition of endosperm proteins. Can. J. Plant Sci. 50: 9-14.

CHEN, C. H., and BUSHUK, W. 1970b. Nature of proteins in triticale and its parental species. II. Gel filtration and disc electrophoresis results. Can. J. Plant Sci. 50: 15-24.

CHEN, C. H., and BUSHUK, W. 1970c. Nature of proteins in triticale and its parental species. III. A comparison of their electrophoretic patterns. Can. J. Plant Sci. 50: 25-30.

COMMITTEE ON ANIMAL NUTRITION. 1964. Agr. Board, Nat. Acad. Sci.–Nat. Res. Counc., U.S.A., and Nat. Comm. Anim. Nutr., Nat. Adv. Comm. Agr. Serv., Canada, Joint U.S.–Can. Tables of Feed Composition. Publ. 1232; p. 118-119. Nat. Acad. Sci.–Nat. Res. Counc., Washington, D.C.

DEDIO, W., SIMMONDS, D. H., HILL, R. D., and SHEALY, H. 1975. Distribution of α-amylase in the triticale kernel during development. Can. J. Plant Physiol. 55:29-36.

DEOSTHALE, Y. G., SURYANARAYANA RAO, K., and MOHAN, V. S. 1969. Nutritive value of some varieties of wheat. J. Nutr. Diet. 6: 182-186.

DE VUYST, A., VERVACK, W., VANBELLE, M., ARNOULD, R., and MOREELS, A. 1958. The amino acid composition of Belgian cereals and their value in feeding pigs and poultry (in French). Agricultura (Louvain) 6: 19-53.

DRONZEK, B. C., ORTH, R. A., and BUSHUK, W. 1974. Scanning electron microscopy studies of triticale and its parental species. In: Triticale: First manmade cereal, ed. by C. C. Tsen; p. 91. Amer. Ass. Cereal Chem.: St. Paul, Minn.

DUBETZ, S. 1972. Effects of nitrogen on yield and protein content of Manitou and Pitic wheats grown under irrigation. Can. J. Plant Sci. 52: 887-890.

EGGUM, B. O. 1968. Aminosyrekoncentration og protein-kvalitet; p. 90. Stongaards Forlag: Copenhagen.

EL GINDY, M. M., LAMB, C. A., and BURRELL, R. C. 1957. Influence of variety, fertilizer treatment and soil on the protein content and mineral composition of wheat, flour and flour fractions. Cereal Chem. 34: 185-195.

ELLIS, R. P. 1971. Electrophoresis of grain proteins: Detection of rye proteins in wheat × rye hybrids. J. Nat. Inst. Agr. Bot. 12: 236-241.

EVERS, A. D. 1970. Development of the endosperm of wheat. Ann. Bot. 34: 547-555.

EVERS, A. D. 1971. Scanning electron microscopy of wheat starch. III. Granule development in the endosperm. Staerke 23: 157-162.

EWART, J. A. D. 1966a. Fingerprinting of glutenin and gliadin. J. Sci. Food Agr. 17: 30-33.

EWART, J. A. D. 1966b. Cereal proteins: Immunological studies. J. Sci. Food Agr. 17: 279-284.

EWART, J. A. D. 1967. Amino acid analyses of cereal flour proteins. J. Sci. Food Agr. 18: 548-552.

FAIRCLOUGH, B. 1947. A microscopical examination of wheat grain. Tech. Educ. Ser., Pamph. 14, 2nd ed. Nat. Joint Ind. Counc. Flour Milling Industry: London.

FEDER, N., and O'BRIEN, T. P. 1968. Plant microtechnique: Some principles and new methods. Amer. J. Bot. 55: 123-142.

FOOD AND AGRICULTURE ORGANIZATION. 1970. Amino acid content of foods and biological data on proteins. FAO Nutr. Stud. 24; p. 285. Food Agr. Organ.: Rome.

GANDHI, S. M., and NOLTRAWAT, K. S. 1968. Influence of nitrogen fertilization on quality characters in a few varieties of common wheat *(Triticum aestivum L.).* Indian J. Agr. Sci. 38: 47-52.

GATES, F. C. 1936. Grasses in Kansas. Rep. Kans. State Board Agr. 55(220-A): 349.

GELMAN, A. L. 1969. Some β-glycosidases

in barley and other cereals. J. Sci. Food Agr. 20: 209-212.

GOLENKOV, V. F., and GILZIN, V. M. 1971. Study of the amino acid composition of the protein of winter rye grain (in Russian). Prikl. Biokhim. Mikrobiol. 7: 328-333.

GONTZEA, I., and SUTZESCU, P. 1968. Natural anti-nutritive substances in foodstuffs and forages. K. Karger, Basle (Switzerland).

GORDON, M. 1922. The development of endosperm in cereals. Proc. Roy. Soc. Victoria. 34: 105-116.

GUNTHARDT, H., and McGINNIS, J. 1957. Effect of nitrogen fertilization on amino acids in whole wheat. J. Nutr. 61: 167-176.

HAY, J. C. 1942. The distribution of phytic acid in wheat and a preliminary study of some of the calcium salts of this acid. Cereal Chem. 19: 326-333.

HECTOR, J. M. 1936. Introduction to the botany of field crops; Vol. 1, Cereals; p. annesburg, South Africa.

HEW, C. L., and UNRAU, A. M. 1970. Investigation of the starch components of a synthetic cereal species. J. Agr. Food Chem. 18: 657-662.

HITCHCOCK, A. S. 1936. The genera of the grasses of the United States. U.S. Dep. Agr. Bull. 772, revised by A. Chase; p. 302.

HITCHCOCK, A. S. 1950. Manual of the grasses of the United States. U.S. Dep. Agr. Misc. Publ. 200: 1051.

HULSE, J. H., and LAING, E. M. 1974. Nutritive value of triticale protein; p. 96. Int. Develop. Res. Cent.: Ottawa, Canada.

HUTCHEON, W. L., and PAUL, E. A. 1966. Control of the protein content of Thatcher wheat by nitrogen fertilization and moisture stress. Can. J. Soil Sci. 46: 101-108.

JACOBSEN, J. V., KNOX, R. B., and PYLIOTIS, N. A. 1971. The structure and composition of aleurone grains in the barley aleurone layer. Planta (Berlin) 101: 189-209.

JANICKI, J., and KOWALCZYK, J. 1965. The biological value of rye proteins compared with wheat protein (in German). Vitalst. Zivilisationskr. 10: 14-21.

JENNINGS, A. C., and MORTON, R. K. 1963. Changes in nucleic acids and other phosphorus containing compounds of developing wheat grain. Aust. J. Biol. Sci. 16: 332-341.

JOHNSON, D. W., and PALMER, L. S. 1934. The nutritive properties of protein, vitamins B and C and the germ in rye. J. Agr. Res. 49: 169-181.

JOHNSON, V. A., and MATTERN, P. J. 1972. Summary report of research findings from improvement of the nutritional quality of wheat through increased protein content and improved amino acid balance. July 1, 1966-Dec. 31, 1972. Ag. Int. Dev., Dep. State: Washington, D.C.

JONES, D. B., CALDWELL, A., and WIDNESS, K. D. 1948. Comparative growth promoting value of the proteins of cereal grains. J. Nutr. 35: 639-650.

KENT, N. L. 1966. Technology of cereals. Pergamon: Oxford.

KENT-JONES, D. W., and AMOS, A. J. 1967. Modern cereal chemistry, 6th ed; p. 730. Food Trade Press Ltd.: London.

KIES, C., and FOX, H. M. 1970a. Effect of level of total nitrogen intake on second limiting amino acid in corn for humans. J. Nutr. 100: 1275-1285.

KIES, C., and FOX, H. M. 1970b. Determination of the first limiting amino acid of wheat and triticale grain for humans. Cereal Chem. 47: 615-625.

KIHLBERG, R., and ERICSON, L. E. 1964. Amino acid composition of rye flour and the influence of amino acid supplementation of rye flour and bread on growth, nitrogen efficiency ratio and liver fat in the growing rat. J. Nutr. 82: 385-394.

KLASSEN, A. J., and HILL, R. D. 1971. Comparison of starch from triticale and its parental species. Cereal Chem. 48: 647-654.

KLASSEN, A. J., HILL, R. D., and LARTER, E. N. 1971. Alpha-amylase activity and carbohydrate content as related to kernel development in Triticale. Crop Sci. 11: 265.

KNIERIEM, W. von. 1900. Der Roggen als Kraftfuttermittel. Landwirt. Jahrb. 29: 483-523.

KNIPFEL, J. E. 1969. Comparative protein quality of triticale, wheat and rye. Cereal Chem. 46: 313-317.

KON, S. K., and MARKUZE, Z. 1931. The biological values of the proteins of breads baked from rye and wheat flours alone or combined with yeasts or soya bean flour. Biochem. J. 25: 1476-1484.

LAPORTE, J., and TREMOLIERES, J. 1962. Inhibitor action of rice, oats, maize, barley, wheat, rye and buckwheat flours on certain proteolytic enzymes of the pancreas (in French). C. R. Soc. Biol. 156: 1261-1263.

LARSEN, I., and NIELSEN, J. D. 1966. The effect of varying nitrogen supplies on the content of amino acids in wheat grain. Plant Soil 24: 299-307.

LAWRENCE, J. M., DAY, K. M., HUEY, E.,

and LEE, B. 1958. Lysine content of wheat varieties, species and related genera. Cereal Chem. 35: 169-178.

LEE, W. Y., and UNRAU, A. M. 1969. Alpha-amylase of a synthetic cereal species. J. Agr. Food Chem. 17: 1306-1311.

LUI, N. S. T., and ALTSCHUL, A. M. 1967. Isolation of globoids from cotton seed aleurone grain. Arch. Biochem. 121: 678-684.

MacAULIFFE, T., and McGINNIS, J. 1971. Effect of antibiotic supplements to diets containing rye for chick growth. Poultry Sci. 50: 1130-1134.

MacMASTERS, M. M., HINTON, J. J. C., and BRADBURY, D. 1971. Microscopic structure and composition of the wheat kernel. In: Wheat: Chemistry and technology, ed. by Y. Pomeranz. Amer. Ass. Cereal Chem.: Set. Paul, Minn.

MacRITCHIE, F. 1972. The fractionation and properties of gluten proteins. J. Macromol. Sci., Chem. A6(4): 823-829.

MADL, R. L., and TSEN, C. C. 1973a. Proteolytic activity of triticale. Cereal Chem. 50: 215-219.

MADL, R. L., and TSEN, C. C. 1973b. Trypsin and chymotrypsin inhibitors of triticale. Abstr., 58th Annu. Mtg., St. Louis, Mo.; p. 78. Amer. Ass. Cereal Chem.: St. Paul, Minn.

MADL, R. L., and TSEN, C. C. 1974. Trypsin and chymotrypsin inhibitors of triticale. In: Triticale: First man-made cereal, ed. by C. C. Tsen; p. 168. Amer. Ass. Cereal Chem.: St. Paul, Minn.

MARKLEY, K. S. 1950. Soybeans and soybean products. Interscience: New York.

MAY, L. H., and BUTTROSE, M. S. 1959. Physiology of cereal grain. II. Starch granule formation in the developing barley kernel. Aust. J. Biol. Sci. 12: 146-159.

McCANCE, R. A., WIDDOWSON, E. M., MORAN, T., PRINGLE, W. J. S., and MACRAE, T. F. 1945. The chemical composition of wheat and rye and of flours derived therefrom. Biochem. J. 39: 213-222.

McDERMOTT, E. E., and PACE, J. 1960. Comparison of the amino acid composition of the protein in flour and endosperm from different types of wheat with particular reference to variation in lysine content. J. Sci. Food Agr. 11: 109-115.

McELROY, L. W., CLANDINNIN, D. R., LOBAY, W., and PETHYBRIDGE, S. I. 1949. Nine essential amino acids in pure varieties of wheat, barley and oats. J. Nutr. 37: 329-336.

McGINNIS, J. 1972. Biological evaluation of cereal grains for nutritional value; triticales, rye, wheat. Rep. submitted to Rockefeller Found. and CIMMYT. Dep. Anim. Sci., Wash. State Univ.: Pullman, Wash.

McLENNON, E. 1920. The endophytic fungus of Lolium. Proc. Roy. Soc. Victoria. 32: 252-255.

McNEAL, F. H., BERG, M. A., McGUIRE, C. F., STEWART, V. R., and BALDRIDGE, D. E. 1972. Grain and plant nitrogen relationships in eight spring wheat crosses, Triticum aestivum L. Crop Sci. 12: 599-602.

MEREDITH, O. B., and WREN, J. J. 1966. Stability of the molecular weight distribution in wheat flour proteins during dough making. J. Sci. Food Agr. 20: 235-237.

MILLER, D. F. 1958. Composition of cereal grains and forages. Publi. 585, Nat. Acad. Sci., Nat. Res. Counc., Washington, D.C.

MITCHELL, H. H., and BLOCK, R. J. 1946. Some relationships between the amino acid contents of proteins and their nutritive values for the rat. J. Biol. Chem. 163: 599-620.

MITCHELL, H. H., and HAMILTON, T. S. 1929. The bio-chemistry of the amino acids. Amer. Chem. Soc. Monogr. Ser. Chemical Catalogue Co. Inc.: New York.

MUNCK, L. 1964. The variation of nutritive value in barley. I. Variety and nitrogen fertilizer effects on chemical composition and laboratory feeding experiments. Hereditas 52: 1-35.

MUNCK, L. 1968. Cereals for feed—quality and utilization. J. Swed. Seed Ass. 78: 137-201.

MUNCK, L. 1969. Genotype environment interaction in protein production and utilization. In: Proc. panel meeting on new approaches to breeding for plant protein improvement; p. 173-186. Int. At. Energy Ag.: Vienna.

MUNCK, L. 1972. Improvement of nutritional value in cereals. Hereditas 72: 1-128.

MÜNTZING, A. 1963. Some recent results from breeding work with rye-wheat in recent plant breeding research. Svaloef, 1946-1961: 167.

OBERMEYER, E. 1916. Untersuchungen uber das Blühen und die Befruchtung von Winterroggen und Winterweizen. Z. Pflanzenzuecht. 4: 347-403.

OSBORNE, T. B. 1907. The proteins of the wheat kernel. Carnegie Inst. Wash. Publ. 84.

PERCIVAL, J. 1921. The wheat plant. Duckworth: London, p. 129-138.

PEREIRA, A. 1963. Studies on wheat protein variation (in Portuguese). Agron. Lusitana 25: 567-582.

PETERSON, R. F. 1965. Wheat. Botany, cultivation and utilization; p. 422. Interscience: New York.

POLANOWSKI, A. 1967. Trypsin inhibitor from rye seeds. Acta Biochim. Pol. 14: 389-395.

POMERANZ, Y. 1973. Structure and mineral composition of cereal aleurone cells as shown by scanning electron microscopy. Cereal Chem. 50: 504-511.

PRIMOST, E. G., RITTMEYER, G., and MAYR, H. H. 1967. Experiments to improve the straw strength of cereals. III. Field experiments with chlor-cholinchloride on winter rye (in German). Bodenkultur 18: 41-56.

REDMAN, D. G., and EWART, J. A. D. 1967. Disulphide interchange in cereal proteins. J. Sci. Food Agr. 18: 520-523.

RUKOSUEV, A. N., and SILANT'EVA, A. G. 1972. Amino acid composition of rye grain, sieved rye flour, bread flour and bran. Vop. Pitan. 31: 142-145.

SALMON, J. 1940. Quelques remarques sur l'etat physique et le comportement histochimique des globoides. C. R. Acad. Sci. Paris. 211: 510.

SANDSTEDT, R. M. 1946. Photomicrographic studies of wheat starch. I. Development of the starch granules. Cereal Chem. 23: 337-359.

SHEALY, H. E., and SIMMONDS, D. H. 1973. The early developmental morphology of the Triticale grain. Proc. 4th Int. Wheat Genet. Symp., Columbia, Mo. p. 265-270.

SHEPHERD, K. W. 1968. Chromosomal control of endosperm proteins in wheat and rye. Proc. 3rd Int. Wheat Genet. Symp., Canberra, Aust. p. 86-96.

SHEPHERD, K. W. 1973. Homoeology of wheat and alien chromosomes controlling endosperm protein phenotypes. Proc. 4th Int. Wheat Genet. Symp. Columbia, Mo. p. 745-760.

SHEPHERD, K. W., and JENNINGS, A. C. 1971. Genetic control of rye endosperm proteins. Experientia 27: 98-99.

SIHLBOM, E. 1962. Amino acid composition of Swedish wheat protein. Acta Agr. Scand. 12: 148-156.

SIMMONDS, D. H. 1962. Variations in the amino acid composition of Australian wheats and flours. Cereal Chem. 39: 445-455.

SIMMONDS, D. H. 1974. The structure of the developing and mature triticale kernel. In: Triticale: First man-made cereal. ed. by C. C. Tsen; p. 105. Amer. Ass. Cereal Chem.: St. Paul, Minn.

SIMMONDS, D. H., BARLOW, K. K., and WRIGLEY, C. W. 1973. The biochemical basis of grain hardness in wheat. Cereal Chem. 50: 553-562.

SIMMONDS, D. H., and ORTH, R. A. 1973. Structure and composition of cereal proteins as related to their potential industrial utilization. In: Industrial uses of cereals, ed. by Y. Pomeranz; p. 51-120. Amer. Ass. Cereal Chem.: St. Paul, Minn.

SOEST, P. J. van. 1966. Non-nutritive residues: A system of analysis for the replacement of crude fiber. J. Ass. Offic. Anal. Chem. 49: 546-551.

SOMIN, V. I. 1970. Amino acids composition of different rye varieties prevalent in the U.S.S.R. (in Russian). Vop. Pitan. 29: 48-54.

SOSULSKI, F. W., PAUL, E. A., and HUTCHEON, W. L. 1963. The influence of soil moisture, nitrogen fertilization and temperature on quality and amino acid composition of Thatcher wheat. Can. J. Soil Sci. 43: 219-228.

SOUTHGATE, D. A. T. 1969. Determination of carbohydrates in foods. I. Available carbohydrates. II. Unavailable carbohydrates. J. Sci. Food Agr. 20: 326-335.

SPURR, A. R. 1969. A low viscosity epoxy resin embedding medium for electron microscopy. J. Ultrastruct. Res. 26:31-43.

SRIVASTAVA, K. N., PRAKASH, V., DE, R., and NAIK, M. S. 1971. Improvement of the quality of proteins of "S-227" wheat by nitrogen fertilization under rain fed conditions. Indian J. Agr. Sci. 41: 202-205.

SURE, B. 1954. Relative nutritive values of proteins in whole wheat and whole rye and effect of amino acid supplements. J. Agr. Food Chem. 2: 1108-1110.

SWAMINATHAN, M. S., AUSTIN, A., KAUL, A. K., and NAIK, M. S. 1969. Genetic and agronomic enrichment of the quantity and quality of proteins in cereals and pulses. In: Proc. panel meeting on new approaches to breeding for plant protein Ag.: Vienna.

TKACHUK, R. 1969. Nitrogen-to-protein conversion factors for cereals and oilseed meals. Cereal Chem. 46: 419-423.

TKACHUK, R., and IRVINE, G. N. 1969. Amino acid compositions of cereals and oilseed meals. Cereal Chem. 46: 206-218.

TRZEBSKA-JESKE, I., and MORKOWSKA-GLUZINSKA, W. 1963. Contents of nitrogen and some essential amino acids in Polish rye grain harvested in 1959 (in Polish). Rocz. Panstw. Zakl. Hig. 14: 153-166.

VIRTANEN, A. I. 1964. Primary plant substances and decomposition reactions in crushed plants, exemplified mainly by studies on organic sulfur compounds in vegetables and fodder plants; p. 3-6. Biochem. Inst.: Helsinki.

WIERINGA, G. W. 1967. On the occurrence of growth inhibiting substances in rye. H. Weenman en Zonen, N. V. Wageningen. p. 68.

WILLIAMS, P. C. 1966. Reasons underlying variations in the protein content of Australian wheat. Cereal Sci. Today 11: 332-335, 338.

WORLD HEALTH ORGANIZATION. 1973. Energy and protein requirements. WHO Tech. Rep. Ser. 522. FAO Nutr. Meetings, Rep. Ser. 52. Rep. Joint FAO/WHO Ad Hoc Expert Comm.

WRIGLEY, C. W. 1970. Protein mapping by combined gel electrofocusing and electrophoresis: Application to the study of genotypic variations in wheat gliadins. Biochem. Genet. 4: 509-516.

WRIGLEY, C. W., and SHEPHERD, K. W. 1973. Electrofocusing of grain proteins from wheat genotypes. Ann. N. Y. Acad. Sci. 209: 154-162.

WRIGLEY, C. W., and SHEPHERD, K. W. 1974. Identification of Australian wheat cultivars by laboratory procedures. Examination of pure samples of grain. Aust. J. Exp. Agr. Anim. Husb. 14:796-804.

YAMPOLSKY, C. 1957. Wheat. Wallerstein Lab. Commun. 20: 343-359.

YONG, F. C., and UNRAU, A. M. 1966. Alien genome combinations and influence on amino acid composition of cereal protein fractions. J. Agr. Food Chem. 14: 8-12.

CHAPTER 5

RYE MILLING

T. A. ROZSA
353 W. Broadway
Winona, MN 55987, U.S.A.

I. HISTORICAL INTRODUCTION

Rye milling technology was identical with wheat milling until about 1870 (Anonymous, 1927). During this period, milling was performed by millstones and was referred to as "flat grinding" (closely set grinding elements followed by some regrinding). In the late 19th century, wheat millers started to adopt the "high-grinding" technique (larger gaps between the grinding elements and more stages of regrinding) to make whiter flours. Roller mills and middlings purifiers were introduced at about the same time.

In simple terms, the object of flat grinding was to make as much flour as possible at each grinder. On the other hand, high grinding uses a multiple grinding system which produces middlings from grain and then flour from the middlings. The "gradual reduction" (half high-grinding) system was introduced as a compromise. It became universal in wheat mills (but not in durum wheat mills) by 1925. Parallel developments occurred in the larger commercial rye mills. The shift was to the production of middlings on the 1, 2, 3, and 4 breaks and flour on the 4, 5, and 6 breaks. More reduction or grinding steps were added gradually to grind the middlings into flour by a gradual reduction. Purifiers were not useful in rye milling, although they were already used extensively in wheat milling.

As late as the 1920's, millstones were still used in Europe to finish the grinding of bran and rye middlings, but shortly thereafter they disappeared. Gradually, rye mills adopted innovations such as feed dusters, finishers in conjunction with the breaks, vibro-sifters, impact mills, and detacheurs. In general, the rye milling process evolved with the wheat milling process.

II. PROPERTIES OF RYE GRAIN

Rye grain can undergo a variety of changes that affect its milling quality. It is susceptible to infection by a parasitic fungus called ergot. The infection produces a black ergot body (sclerotia) which is toxic and must be removed from the rye before it is milled or fed to livestock. Rye grain has a strong tendency to sprout in a warm, wet climate. During sprouting, the activity of α-amylase can increase to the extent that the flour cannot be used for bread production. The miller must select rye grain with minimum sprout damage (Shands, 1969). Sprout damage can usually be detected visually by the presence of rootlets, but occasionally it can only be detected by measurements of enzyme activity by tests such as the amylograph or the falling number. Rye grown in different countries varies widely in 1000-kernel weight and the miller must take account of this variability, especially when milling a mixed grist.

In selecting rye for milling, the miller takes into consideration differences in bran thickness. Grain with thin bran gives higher flour yields. Rye grain varies widely in color. Light brown grain is preferred; blue and green grain produces a dirty gray flour. Rye has a higher pentosan content than wheat. Accordingly, it is more difficult to grind. The higher pentosan content makes rye flour extremely hygroscopic. This tends to agglomerate the flour, which interferes with sifting. For this reason, rye is milled at a lower moisture than wheat.

The friction of the rye kernel surface is greater than that of wheat. Rye mills use more pitch for spouting. Terminal velocity of rye in vertical air currents of aspirators is less than that of wheat; about 1640–1960 ft/min (8.3–9.9 m/sec). Rye kernels do not have a hyalin layer outside of aleurone cells, and water penetration into the kernel is swift. Because of this, rye requires shorter temper time than wheat.

What does the U.S. miller look for in milling rye? First, the commercial grade (which includes ergot content, if any), next, the price, and finally, the thin-grain content. The grain sample is then visually examined for soundness and maturity. The German rye miller assesses cleanliness, test weight, color, sprout damage, thickness of the bran, and moisture content.

In the U.S.S.R., the rye miller grinds what is delivered to the mill (Kupric, 1954). The rye is sorted into three classes according to growth habit, vitreousness, hardness, and color. Winter rye is Class I. This class is divided into five subclasses according to region of growth. Class II rye is hard spring with two subclasses. Class III rye is other spring rye with no subclasses. For grading, the following factors are considered: condition = color, odor, taste, moisture, foreign matter, and damaged kernels; milling value = vitreousness, kernel-size uniformity (thin kernels through 1.4×20-mm screen openings), ash content, test weight, 1000-kernel weight, and the laboratory milling test; and baking value = sprouting, rheological properties, and baking test results.

III. RYE MILLING

The milling of rye will be discussed in three subsections: rye milling in North America; rye milling in western Europe; and rye milling in eastern Europe, including the U.S.S.R. North American rye milling is characterized by grinding rather high-protein, more vitreous, and smaller-sized kernels into fine flour with no strict quality requirements. Western European rye milling is characterized by grinding rather low-protein, nonvitreous, and larger-sized kernels into fine flour with strict quality requirements. Eastern European rye mills grind all types of rye kernels into a coarse flour according to local quality specifications.

A. Rye Milling in North America

STORAGE

On U.S. and Canadian farms, rye may be stored in wood structures or, more commonly, in round or square metal bins (Shands, 1969). Commercial storage bins are constructed from concrete and metal. A satisfactory grain storage facility should retain grain quality, exclude water, protect against thieves, rodents, birds, insects, objectionable odors, fire and wind, provide for effective fumigation, and allow for easy sampling. Since rye is similar to wheat in storage requirements, satisfactory storage conditions for wheat are usually satisfactory for rye. To avoid spoilage, the moisture content of rye should not be above 13%.

SELECTION OF THE MILL MIX

On receipt at the mill, the lot of grain is stored according to origin, grade, thin-kernel content, moisture content, ergot content, and degree of sprouting. The mills select rye with less than 8% thin kernels, which pass through 0.064 × 0.375-in. (1.6 × 9.5-mm) screen perforations. The quality of the rye is usually assessed by standard cereal laboratory techniques. Grain for the mill grist is usually selected to yield flour of required specifications from the lowest-cost grist. Assessment of rye quality by test milling is not necessary in most rye mills. Uniformity in the mill mix is most important; since it is attained by blending rye from as many storage bins as possible, it is necessary to know the quality indicators of the grain in each bin.

CLEANING AND TEMPERING

The rye cleaning process is similar to the process used for wheat, with adjustments to accommodate the different shape and size of the rye grain (Shaw, 1970). Wheat cleaning machines are generally used. These include the magnet, milling separator, disk machines, entoleter, scourer, aspirators, and stoners. A schematic diagram of a typical North American rye cleaning process is shown in Figure 1.

The top screen of the milling separator should have slots of 9/64 in. (3.8 mm) wide and 3/4 in. (19 mm) long which should be lined up in the direction of the grain movement. This screen will remove soybeans which, if ground, paste up the roll corrugations. The sand screen should have triangular-shaped perforations of 9/64 in. (3.8 mm) on each side. These will remove buckwheat and cockle seeds. Since rye kernels have a different shape than wheat, the disks in the disk machines for the two grains will be different. The primary rye disk grader should have a set of V6, AC, and X pockets. Head-end liftings of the disk grader go to a recleaner, tail-end liftings are recirculated, and overs go to the oats disk machine equipped with MM pockets to tail over oats and some of the ergot. The ergot rejects can be reclassified on a spiral separator to concentrate the ergot for sale to the pharmaceutical industry. The recleaner should be equipped with V4.5, V5, and V5.5 pockets. The head-end liftings of the recleaner are screenings, tail-end liftings are recirculated, and overs are the milling rye. The clean, dry rye passes through an entoleter scourer-aspirator, operating at 1750 rpm with 2 rows of 12 pegs on 10 3/4-in. (27.3 cm) diameter rotors, to eliminate live insects and remove

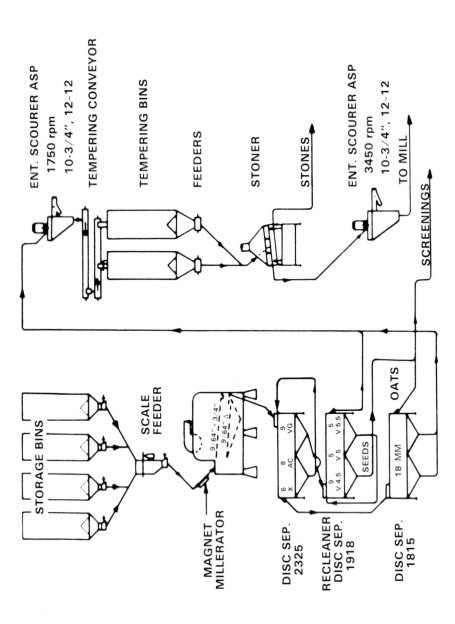

Figure 1. Rye cleaning process diagram for a typical North American rye mill (1200 cwt/24 hr).

loose dirt and bran. The grain is now ready for tempering. After tempering to 14.5–15% moisture for about 6 hr, the grain passes over a stoner and another entoleter scourer-aspirator running at 3450 rpm with the same type of rotor as in the dry-grain entoleter. The liftings from the second entoleter contain germ, bran particles, and insect fragments.

Simple tempering of rye with water is usually sufficient; water penetrates rye grain faster than it penetrates wheat. Rye is milled at 1% less moisture than wheat; 14.5–15–15.5% moisture is preferred, depending on the softness and vitreousness. Short temper times of 6 to 15 hr are generally used in North American rye mills. In the winter, warm tempering water (54°–82°C) brings the grain to the first break at a more uniform temperature, and thereby helps to maintain a more uniform milling operation, eliminates condensation in the break sifters and spouts, and obviates the need for frequent changing of sifter cloths in the head-end break sifters.

RYE MILLING PROCESS

A typical North American rye milling process is shown in Figure 2 (Shaw, 1970). It consists of 5 breaks, 1 bran duster, 1 sizings roll, 7 reduction rolls, and 1 tailings roll. All rolls are corrugated, starting with 14 (5.5/cm) Getchell cuts on first break and increasing to 32 (12.5/cm) Getchell cuts on the fifth break and first sizings rolls. The reduction and tailings rolls have 36 (14/cm) Twin City cuts.

FROM TEMPER

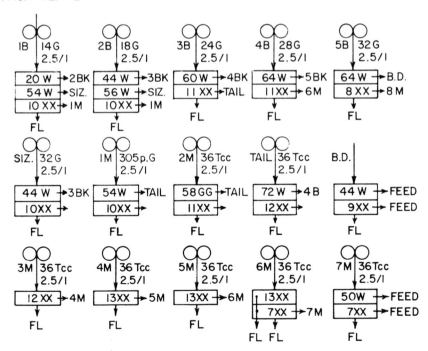

Figure 2. Milling process diagram of a typical North American rye mill (1200 cwt/24 hr).

The reason why smooth rolls are not used in the reduction of rye middlings is because they flake the middlings (because of high pentosan content) which would be subsequently removed as scalpings in the sifting operations to millfeed. All rolls run at a differential of 2.5:1. The total roll surface is about 0.5 in./cwt (2.8 cm/100 kg or 22.5 cm/ton grain) per 24 hr. White rye flour is sifted through bolting cloths ranging from 10XX to 13XX. For dark rye flour, 7XX to 9XX cloths are used. The break rolls should equal about 45% of the total roll surface. Scalping and grading sifter surface should be about 27%, and flour sifting would give the remaining 73%. Total sifter surface of the mill in Figure 2 would be about 0.5 ft^2/cwt (0.1 m^2/100 kg or 0.8 m^2/ton grain) per 24 hr without the rebolt sifters. The maximum load on the first break rolls should not exceed 1.7 lb/min/in. (0.30 kg/min/cm).

With forced grinding, high extraction is aimed for on every roll. Typical North American break releases are given in Table I. Table II represents the quantity and ash content of flour streams from the sifters in the process shown in Figure 2.

In general, total flour extraction varies quite widely from mill to mill. Higher ash specifications and coarser flour silks yield higher extractions. Average flour yield, based on dirty grain, is about 100 lb/2.15 bu of rye or 83% extraction.

TABLE I
North American rye mill break releases

Break	Release %
1	30 through 28 wire
2	30 through 44 wire
3	55 through 54 wire
4	50 through 72 wire
5	50 through 72 wire

TABLE II
North American rye mill flour streams

Stream	Amount %	Ash (14% mb) %
Break		
1	15.62	0.346
2	6.79	0.508
3	4.15	0.925
4	3.36	1.539
5	7.58	2.430
Sizing	16.32	0.370
Reduction		
1	15.00	0.585
2	7.91	0.745
3	2.44	0.947
4	1.58	1.036
5	1.45	1.155
6 (Head)	1.98	1.896
6 (Tail)	8.24	2.200
7	4.94	3.330
Tailings	2.31	1.436
Bran duster	0.33	1.920

The current trend is to use feed dusters on the break scalps before they pass to the next break roll. This helps to increase the yield of white flour.

The pitch of spouts in rye mills should be 50° or higher because of the high friction of rye mill products. The high-release forced grinding used in rye milling generates considerable heat; an efficient exhaust system is required to remove this heat.

RYE FLOUR TREATMENT

The only rye flour treatment used in North America is a light addition of chlorine (0.19 g to 0.31 g/kg or 0.3 to 0.5 oz/cwt) to white rye flour for color improvement.

RYE MEAL MILLING

Most rye mills have a simple 1-, 2-, or 3-pair roll plant to make rye meals (Shaw, 1970). The rolls have 14 to 18 corrugations/in. (5.5 to 7.0/cm) and are run at 2.5:1 differentials. The sifters segregate the chop into three streams on No. 10 and No. 24 mill screen wires. The desired stream is bagged off, while the other separations are routed into the rye flour mill (first break sifter) for finishing off. Clean, dry rye is ground for most meals. For "flaky meals," tempered rye is used.

MILL QUALITY CONTROL

Quality control in a North American rye mill begins with selection of the rye for purchase. This is not very complicated because only a minor proportion of the total crop is purchased for milling and the origin as well as the grade of the grain is known. Prior to milling, binned rye is analyzed for moisture and diastatic activity.

There are no standards of identity for rye flour in the U.S. or in Canada. Accordingly, each rye mill must establish the specifications of its commercial grades of flour. Thereafter, the laboratory checks the uniformity of the flour products by standard cereal laboratory tests. It is common practice to sample the products every 4 hr during milling. The common quality control tests applied to rye flour in North America are listed in Table III. The baking quality test is very important and must be done with the proper equipment. If the rye flour is tested in a blend with wheat flour, the wheat flour used must be of uniform quality. In testing rye flour, it is always essential to follow commercial baking conditions as closely as possible. Commercially milled flours can be duplicated in the

TABLE III
North American rye mill quality control tests

Test	AACC Approved Method No.
Moisture	44-19
Ash	8-01
	8-02
Color (Pekar)	14-10
Diastatic activity	
Brabender Amylograph®	22-10
Hagberg falling number	56-81B
Extraneous matter	
White rye flour: Acid hydrolysis method	28-40
Other: Tween-Versene method	28-60
Baking quality	10-20

laboratory. It is desirable to be able to mill a 0.85–1.0 ash straight-grade flour (Zwingelberg and Reimers, 1972). This can be done by the Buhler laboratory mill: 1) it can be equipped with corrugated reduction rolls, and the overs of these passed twice through a laboratory (Buhler) bran finisher to make 82% extraction; and 2) if the smooth reduction rolls are retained (60% extraction), the overs should pass through a laboratory hammer mill with 2/64-in. screen openings, and then through the laboratory bran finisher with 200-μ screen openings. The short temper time (2 hr) works as well as the longer time (18 hr).

CHARACTERISTICS AND USES OF MILLED PRODUCTS

Flour. Two basic grades of rye flour are generally produced in the U.S. One is the white or light rye flour, the other is the dark rye flour. Normally, the white flour forms about 80% of all the flour made. Many different grades of rye flour are made commercially by combining the two basic grades in various proportions. The blend of all of the two basic grades yields a straight-grade flour and is called medium rye flour.

During growth, the milling properties of the rye grain are influenced greatly by the environment, through its effect on the mealiness and toughness of the kernels. These physical properties are related to the viscosity of the gums of rye grain (Drews, 1965). A high initial (amylograph) viscosity indicates toughness (vitreousness) of the kernels. A low viscosity is usually obtained for flour from mealy (nonvitreous) rye endosperm which gives a higher milling extraction than the vitreous rye endosperm. The flour from mealy rye is usually coarser and has a higher pentosan content than the flour from vitreous endosperm. Starch damage is generally not as high in rye flours as in wheat flours; the gums in rye endosperm seem to protect the starch granules from mechanical damage. Furthermore, the use of corrugated rolls in grinding rye does not damage the starch granules as readily as the smooth reduction rolls do in wheat milling. It is extremely difficult to mill rye flour to a controlled starch damage (Drews and Zwingelberg, 1975). However, recently it was indicated and claimed that the starch damage does not seem to affect the baking quality of rye flours (Vorwerck, 1976).

In the absence of standards of identity, the ash content of North American commercial rye flours varies widely for equivalent flour grades. White rye flours vary from 0.60 to 0.70% in ash content. Dark rye flours are in the range of 2.25 to 3.00% ash. Moisture content is only important relative to milling. Protein content is not an important quality factor in rye flour. Lighter colored rye flour contributes less rye flavor to baked goods than the darker flours.

The principal uses of rye flours are in the production of bread, crackers, snack foods, and prepared flour mixes. Rye breads are popular on the eastern coast of the U.S. and in some metropolitan areas. U.S. rye breads are usually made from rye-wheat flour blends containing up to 80% wheat flour. Cracker and biscuit bakers use some of the less expensive (than wheat) rye flours to improve the quality and lower the cost of their products. Meat packers and processors use some rye flours as fillers and binders in sausages.

Rye Meals. Rye meals may be flaky (coarse, or fine), coarse, medium fine, extra fine, or steel-cut. The meals are used as ingredients of specialty breads, like pumpernickel, unleavened flat rye crisp, snack foods, crackers, and Swedish rye crisps.

Milling By-Products. Rye milling by-products are generally used in the production of formulated animal feeds.

B. Rye Milling in Western Europe

SELECTING, CLEANING, AND TEMPERING

In Europe, rye quality testing is more critical than in North America because the crops vary greatly from year to year. Seed grain is scrupulously cleaned of all foreign matter on the farm prior to seeding or selling (Flechsig, 1955). A medium-size western-European mill would clean the rye (after it passed through an automatic scale) on a milling separator, magnet, four trieurs (oat and cockle, both recleaned), a spiral separator to remove ergot, and a centrifugal sifter for screenings. (Cylindrical trieurs are more popular cleaning machines than the disk machines used in North America.) The grain then goes through the temper, magnet, scourer, brush machine, automatic scale, and second temper spray, and into the grinding bin. Before first break, the grain usually passes through a set of crushing rolls followed by sifters. Some millers advocate washing and/or warm-conditioning the grain, but this is rarely done.

Until about 20 years ago, approximately two-thirds of the West German rye mills employed sulfur dioxide gas or sodium bisulfite (150 g in 0.66 liters of water for 100 kg of grain) to bleach the grain in the first temper conveyor. In addition to bleaching, this treatment sterilizes the grain. Rye bran is thick and, at the necessarily low grinding moistures, it pulverizes and contaminates the color of the best and whitest flour streams. The blue-green bran powder gives a grayish hue to the flour. If the bran is bleached to yellow, the resulting flour has a pleasant creamy color. By using bleach, West German millers obtained 2 to 3% more white rye flour. Bleaching also lightens the color of dark rye flour. In the U.S., bakers prefer unbleached dark rye flour (as-is). The bleaching compounds used do not enter into the rye endosperm, so there is no effect on the functional properties of the flour. The small amount of residue in mill by-products creates no difficulties in mixed animal feeds.

In western European rye milling, the first temper time is usually about 5–6 hr (Kent, 1966). The scourers are usually run at 3100–3600 ft/min (16–18 m/sec) periferal velocity, somewhat higher than that used for wheat cleaning. Even at the higher velocity, the retention time in the scourers will be longer for rye, because the bran of rye offers more friction than the bran of wheat. Rye germ is smaller than that of wheat and can be easily detached in the scourers. The brush end of the rye kernel is broad, and the germ end is pointed with the germ exposed. A screen with a No. 12 mill screen wire sieve can scalp out the germ from the dusty screenings of the fine scourer.

THE MILLING PROCESS

Until 1950, the progress of rye milling technology in Europe was very slow (Hopf, 1939; 1950). Since that time, a number of useful innovations have been introduced. The break rolls have shallower and duller sawtooth corrugations than wheat milling rolls, to release more flour in the breaks. European corrugating machines cut each groove one at a time. The grooves have two straight sides with a narrow flat surface on the top. The two sides are specified by the angle they form with the radial direction. Frequently, the slow roll will have one to two grooves/cm (2 to 5/in.) more than the fast roll. When both rolls have the same number of corrugations, it is possible to interchange the worn fast roll with the still sharp slow roll. Until recently, European roller mills operated at almost half the periferal speed compared with U.S. mills. The same crushing and cutting action is achieved by doubling the spiral of the grooves. Differentials are similar in western European and U.S. rye mills, up to 3:1 on corrugated rolls.

Typical western European rye milling starts with smooth crushing rolls (omitted in new mills), followed by five to seven breaks, two or more bran finishers, four to six reduction rolls, and one or two tailings rolls. In the reductions, four to six flake buster detacheurs are frequently used ahead of the sifters. All rolls, except the crushing rolls, are corrugated. Rebolt sifters are standard parts of the process. Twenty-five centimeter (10-in.) rolls with speeds used in the U.S. can average 0.4 in./cwt (2.2 cm/100 kg or 18.5 cm/ton grain) per 24 hr. A "square" sifter bolting surface of 0.45 ft^2/cwt (0.09 m^2/100 kg or 0.75 m^2/ton grain) per 24 hr can be used. The square sifters are relatively new; most mills still use plan sifters.

In modern European rye milling, 12-in. (30-cm) diameter rolls are recommended with spirals of 20% (2.5 in./ft) (Rada, 1955). Flour is sifted on No.'s 8 to 12 silk cloths. Currently, there is a trend to run the rolls dull to dull instead of sharp to sharp as was the case earlier. For grain with large kernels, the head-end breaks are normally set sharp to dull. The slope of the corrugations depends on the vitreousness of the rye kernels; mealy rye requires 5% steeper angles than hard vitreous grain. There is a gradual introduction of bran and shorts finishers in rye milling in western Europe. These machines process the scalps of first-, second-, and third-break sifters and may have their own sifter sections with finer silks. Impact mills are included in some mills to grind the overs of the first reduction sifter.

There has been a great deal of European work on the properties of millstream flours (Pelshenke *et al.*, 1963; Drews and Reimer, 1965; Drews, 1965, 1966; Weipert, 1970). Some of this work is discussed in **Chapter 6**.

Rye milling in western Europe, particularly West Germany, has been further improved by the success of the combination mills built to grind both wheat and rye. Many western European rye mills are not fully automatic, but these will not be discussed.

C. Rye Milling in Eastern Europe and in the U.S.S.R.

SELECTING, CLEANING, AND TEMPERING

In eastern European countries, including the U.S.S.R., the mill mix is selected on the basis of uniformity of kernel size and baking properties (α-amylase activity). Milling properties, like plump kernels, proportion of thin kernels, test weight, grain color, and moisture, are secondary considerations. Usually the grain had not been inspected before it arrived at the mill. The grain is graded in the mill laboratory and binned according to grade and moisture content. After the rye is allocated in the elevator and turned over, grain in each bin is tested in the mill laboratory for properties such as test weight, color, ash, kernel size, and α-amylase activity.

Rye cleaning in eastern Europe is similar to wheat cleaning and to North American rye cleaning (Kupric, 1954). One minor difference is that the scourers are run at higher speeds. A typical cleaning process starts with an automatic scale, followed by milling separator (10 mm ϕ, 6.5 mm ϕ, 1.25 × 20-mm screens), magnet, trieurs (cockle and oats, with recleaners for both), magnet, scourer, separator, magnet, first temper (6–10 hr), magnet, scourer, separator, and second temper (1.25 hr). The tempered grain then goes to the smooth crushing rolls, both run at the same speed. The screenings are sifted on No.'s 16 and 10 mill screen wires and the fines are aspirated before being ground to rye mill feed.

THE MILLING PROCESS

Rye milling technology in the U.S.S.R. is highly developed (Kupric, 1954). Break-roll corrugations run from 5 to 9 cuts/cm (12.5 to 23/in.), with spirals of 12.5–15% (1.5–1.75 in./ft). The break rolls run sharp to sharp and the reduction rolls dull to dull. The sifter cloths are somewhat coarser than in the U.S. practice. Some mills have U.S.-type bran dusters to process the overs of the last break sifter. Centrifugal sifters are included after bran and shorts dusters.

The more vitreous the grain, the greater the number of grinding steps used. The U.S.S.R. roller mills are normally driven at 332 m (1090 ft) per min (430 rpm) periferal velocity compared with 396 m (1300 ft) per min (510 rpm) in the U.S., and 183 m (600 ft) per min (230 rpm) in western Europe. The characteristics of their technology are: coarser flour granulation, darker flour color, and great thruput of their plants. Therefore, they use coarser flour sifting cloths and coarser and deeper roll corrugations.

Based on the technology described by Kupric (1954), the U.S.S.R. Ministry of Bread Products (Anonymous, 1958) has established norms (standards) on the construction and operation of rye mills. Prescribed production norms for U.S.S.R. rye mills are presented in Table IV. To accomplish the prescribed norms, process A requires six or seven breaks and six or seven reductions, process B requires four or five breaks and three to five reductions, and process C requires four or five breaks and one or two reductions. Capacity specifications for the three processes are given in Table V. The characteristics of the main rye flours produced in the U.S.S.R. are given in Table VI.

TABLE IV

Yield of products for three types of U.S.S.R. rye mills

Process Type	A %	B %	C %
Flour I	63	15	...
Flour II	...	65	87
Flour III	15	...	...
Rye middlings (feed)	18	16	9
Screenings	2.8	2.8	2.8
Losses	1.2	1.2	1.2

TABLE V

U.S.S.R. rye mill equipment norms

Process	Specific Roll Surface/24 hr			Plan Sifter Surface[a]/24 hr		
	in./cwt	cm/100 kg	cm/ton grain	ft²/cwt	m²/100 kg	m²/ton grain
A	0.325	1.82	14.40	0.97	0.20	1.60
B	0.163	0.91	7.35	0.44	0.09	0.72
C	0.135	0.76	6.00	0.38	0.08	0.62

[a]Not square sifter.

Figure 3 is a typical process diagram of a process A, U.S.S.R. rye mill designed to grind 100 tons (1700 cwt) per 24 hr. It has 14 m (550 in.) of roll length, two bran finishers, two shorts dusters, with specific roll surface of 1.9 cm/100 kg or 15 cm/ton grain (0.34 in./cwt) per 24 hr. The "plan sifter" surface is 167 m² (1800 ft²) including two centrifugals and rebolt sifters. For all sifters, the specific surface is 0.22 m²/100 kg or 1.76 m²/ton grain (1.1 ft²/cwt) per 24 hr. The process diagram shows crushing rolls, five breaks, and seven reductions. All rolls are 25 cm (10 in.) in diameter, and are run at 360 m (1180 ft) per min periferal velocity (450 rpm). The crushing roll operates at 1.5:1 differential and so do the No.'s 5, 6, and 7 reduction rolls. The remaining rolls operate at 2.5:1 differential. The break rolls have 25°/65° sawtooth cuts and the reduction rolls 40°/70°. The first three breaks grind with corrugations set sharp to sharp; all other rolls are set dull to dull. Figure 3 also tabulates the grinding roll specifications.

The flour silks for grade I (1 FL in Figure 3) flour are No.'s 7, 8, and 9; rebolts use No.'s 7 and 8, and for grade III (III FL in Figure 3) flour, the silks are No.'s 8, 9, and 10, and their rebolts use No.'s 8 and 9 silks. Generally, the first three breaks feed the first three reductions which in turn form the stock for the next three reductions. The seventh reduction finishes the leftover stock from the sixth reduction and the fifth break.

Recent publications indicate that the trend in the U.S.S.R. is to increase plant capacity and extraction and improve flour quality. The innovations that are being introduced include dehulling before tempering, centrifugal break scalp sifters, finer silks for Grade I flour on the head-end, and coarser silks for Grades II and III flours on the tail-end of the mills.

IV. AIR CLASSIFICATION OF RYE FLOUR

The type of products that can be obtained from rye flour can be dramatically increased by air classification. In the U.S., the Pillsbury Company pioneered the application of air separators to cereal flours and discovered the so-called "protein shifting" (Rozsa *et al.*, 1963a). The patent covers commercial processes for the air classification of rye flour, white rye flour, and dark rye flour. Fine grinding of the cereal flours, including rye flours, prior to air classification, improves the efficiency of the process (Rozsa *et al.*, 1963b). The air-classified rye flour fractions show greatly different protein, ash, maltose, and water-absorption values. Rye flours from soft and semi-hard grains classify easily to produce a fine-particle-size fraction with more than 20% protein and an intermediate-particle-size fraction with less than 3% protein.

Considerable work on air classification of rye flour has been carried out in West Germany on pilot-plant scale (Cleve, 1960; Drews, 1965; Drews and

TABLE VI
Specifications of U.S.S.R. rye flour grades

	% Ash (14% mb)	Sieve/ % Overs	Sieve/ % Thrus	Color
I grade	0.65	No. 5/2	No. IX/90	White
II grade	1.25	450 μ/2	No. IX/60	Gray
Straight-grade	1.70	670 μ/2	No. IX/30	Specky

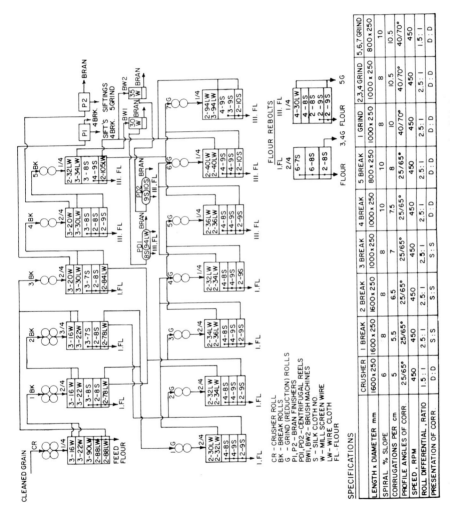

Figure 3. Process diagram and specifications of a U.S.S.R. process A rye mill (1700 cwt or 100 tons grain/24 hr).

Reimer, 1970). The German workers have shown that extensive shifts can be achieved in pentosan content and α-amylase activity of the possible fractions. Both the pentosans and the α-amylase shifted toward the coarsest fraction. The intermediate-particle-size fraction showed the highest amylogram peak values.

Research on air classification of rye flour on a laboratory scale has been carried out by Anderson and coworkers (1972; 1974). The protein shifting potential of rye flour was compared with that of wheat and triticale flours.

As yet, air classification of rye flours has not been applied commercially. The main reasons for this are the small market for the products obtained, the high power requirement for the process, and the considerable loss of moisture in the process.

V. POTENTIALS FOR RYE FLOURS

Low-protein, air-classified, or head-end rye flours could have some advantages as brewing adjuncts, but this application has not yet been commercially accepted. Rye starch is highly susceptible to amylase attack (fast conversion) and the soluble proteins could be good yeast foods and foam builders. Rye flour pentosan is a good foam stabilizer.

The low-protein, air-classified fraction might be useful in chocolate and darker prepared flour premixes. White rye flour can be combined with complimentary anticaking agents to produce a suitable bakery dusting flour which can be used instead of corn starch.

A considerable quantity of rye flour could be used in the U.S. for the manufacture of pet foods. Rye starch is readily digestible, and this property is important in the use of rye flour for the production of food for pets like dogs and cats. Rye clear flours could have high protein content (17–18%) and this protein is of higher nutritive quality than wheat protein. Rye pentosans have a high water-absorbing capacity; another useful functional property for pet-food manufacturing. The color of the flour is not a factor, since most pet foods are dark.

From the wheat-flour miller's viewpoint, the current dark rye flour is a mixture of rye "clears with red dog." For most industrial markets, a rye clear without the red dog would be preferable. No change is needed in the milling; a new dark rye flour rebolt sifter would take care of everything. Over 9XX stock would go to rye feeds. Between 9XX and 11XX, the stock meets original dark rye specifications. The throughs of 11XX we could call "rye clear," with about 2% ash and 17% protein content. This rye clear could be commercially improved by a sulfur dioxide (reducing) bleach or by a hydrogen peroxide spray (oxidizing) in an agitator, depending on the end use.

Rye clear flours (not dark rye) could be used to produce "composite flours" which can be used to prepare a variety of products. These would have the flavor of rye preferred by some people and a slightly higher nutritional quality than similar products made from 100% wheat. The nutritionally limiting amino acids of the rye clear flour proteins are lysine and isoleucine[1].

Laboratory and commercial tests have proven the value of soluble and insoluble gums of rye flours as substitutes for other gums (guar, locust, etc.) as

---

[1]T. A. Rozsa, 353 W. Broadway, Winona, MN 55987; unpublished data.

wet-end additives in paper manufacture. The low gelatinization temperature of rye starch permits the use of higher machine speeds than are used for corn starch. Some grades of rye flour are used as foundry core binders, ore pellet binders (may be precooked), and plywood glue extenders.

There are many potential food and nonfood applications where rye flour and/or its components can be conveniently used because of the characteristic functional properties of the rye endosperm.

LITERATURE CITED

ANDERSON, R. A., STRINGFELLOW, A. C., and GRIFFIN, JR., E. L. 1972. Preliminary processing studies reveal triticale properties. Northwest. Miller 279: 10-13.

ANDERSON, R. A., STRINGFELLOW, A. C., WALL, J. S., and GRIFFIN, JR., E. L. 1974. Milling characteristics of triticale. Food Technol. 28: 11, 66-76.

ANONYMOUS, 1927. Taschenbuch des Müllers (8th ed.). MIAG CO.: Braunschweig. pp. 20-88.

ANONYMOUS, 1958. The ratification of rules for organization and management of the technological process in the mills of the ministry of bread products. U.S.S.R. - Order No. 409. (Translated from Russian to English.)

CLEVE, H. 1960. Möglichkeiten der Beeinflussung der Mehlqualität durch Wind Sichtung. Getreide Mühle 4: 141-157.

DREWS, E., and REIMER, H. 1965. Einige Qualitätsmerkmale von Roggenpassagenmehlen in Abhängigkeit vom Diagram. Die Mühle 102: 67-70.

DREWS, E. 1965. Zusammensetzung und Eigenschaften durch Windsichtung erhaltener fraktionen bei Roggenmehl. Getreide Mehl 15: 129.

DREWS, E. 1966. Die Bedeutung der Schleimstoffe für die Bewertung der Roggenqualität. Getreide Mehl 16: 21-26.

DREWS, E., and REIMER, H. 1970. Ergebnisse der Windsichtung von Inlandroggen Mehlen. Die Mühle 107: 598-599.

DREWS, E., and ZWINGELBERG, H. 1975. Die Verarbeitungseigenschaften von Roggen unterschiedlicher Kornbeschaffenheit unter Berüksichtung der granulation der Mehle. Die Mühle 112: 595.

FLECHSIG, J. 1955. Fachkunde für Müller. Moritz Schafer, Detmold.

HOPF, L. 1939. Taschenbuch für Müller und Mühlenbau, 2nd edition. Moritz Schafer, Detmold.

HOPF, L. 1950. Mühlen Technisches Practikum, Band I. Hugo Mathes, Stuttgart.

KENT, N. L. 1966. Technology of cereals. Pergamon Press: London.

KUPRIC, Ja. N. 1954. Malomipari Technologia. Elelmiszeripari es Begyüjteski Könyv es lapkiadó, Budapest. (Translation from Russian to Hungarian).

PELSHENKE, P. F., BOLLING, H., and ZWINGELBERG, H. 1963. Studien über die Roggenmehl qualtität im praktischem Mühlenbetrieb. Die Mühle 100: 537-539, 547-555.

RADA, I. T. 1955. Örlö es Hantoloiparok - Tankönyvkiado, Budapest.

ROZSA, T. A., HARREL, C. G., MANNING, W. T., WARD, A. B., and GRACZA, R. 1963a. Cereal flour fractionation process. U.S. Patent 3,077,408 (filed November 1954).

ROZSA, T. A., WARD, A. B., and GRACZA, R. 1963b. Process of reducing and surface treating cereal endosperm particles and production of new products through attendant separations. U.S. Patent 3,077,308 (filed March 1956).

SHANDS, H. L. 1969. Cereal science. AVI Pub. Co., pp. 118-149.

SHAW, M. 1970. Rye milling in U.S.A. - Assoc. of Operative Millers, Bull. pp. 3203-3207.

VORWERCK, K. 1976. Der Einfluss unterschiedlicher Stärkebeschädigung beider Roggenvermahlung auf die Analysen und Backergebnisse. Getreide Mehl Brot 30(1): 1-5.

WEIPERT, O. 1970. Beeinflussung einiger Qualitätsmerkmale des Roggens durch Trocknungsmassnahmen. Die Mühle 107: 424-426, 437-439, 450.

ZWINGELBERG, H., and REIMERS, H. 1972. Herstellung von Roggenversuchsmehlen. Die Müehle 109: 773-774.

CHAPTER 6

BREAD-BAKING AND OTHER USES AROUND THE WORLD

E. DREWS
W. SEIBEL
Federal Research Institute of Cereal and Potatoe Processing
Detmold, Federal Republic of Germany

I. INTRODUCTION

A. General

The history of rye bread production is as old as rye growing itself. However, there is a remarkable difference between the bread food of the ancient people and our bread. The production of rye bread is no longer a question of chance, but a planned, highly technical process. Much research has been done; the optimal conditions of the main phases of bread production, *i.e.*, mixing, fermentation, and baking, are known. The oldest, and still the most important, fermentation process in rye bread production uses sour dough. The microbiology of sour dough is well understood and will be discussed in this chapter.

The principal difference between wheat flour and rye flour is that rye proteins cannot form gluten after mixing with water, which is the basis of wheat bread production. Other substances in the rye, however, are able to bind the water during mixing to produce a dough that can be baked into bread. These are the pentosans. They will be discussed in some detail.

Only in a few countries is rye considered as a bread grain. The major ones are Austria, Czechoslovakia (C.S.S.R.), Federal Republic of Germany, the German Democratic Republic (G.D.R.), Poland, and the Soviet Union (U.S.S.R.). In the other rye-producing countries, it is used mainly as a feed grain.

B. Grading of Rye and Baking Strength

Only a few countries have special rye classification or grading systems. All known rye grading systems are based mainly on external characteristics of the grain. Of the biochemical characteristics, only the α-amylase activity is included in some grading systems.

CANADA

The Canadian grading system (Canadian Wheat Board, 1974) consists of four standard grades and one special grade for ergoty rye (Table I). For milling and baking purposes, grades No. 1 and 2 Canada Western (C.W.) are generally used. Occasionally No. 3 C.W. rye is milled for bread. Grade No. 4 rye usually has high percentages of heat-damaged kernels, and therefore the milling and baking qualities of this grain are very poor. Heat-damaged kernels are produced by self heating (due to excessive moisture in the grain) or by artificial drying at high temperatures. The high temperatures alter the native properties of the proteins and pentosans and thereby decrease the water absorption of the flour. Removal of ergot from the special grade by cleaning is very difficult. Accordingly, ergoty rye is seldom used for milling. In some countries where rye is used for bread, national food laws do not permit the use of rye with ergot contamination higher than 0.2%.

THE UNITED STATES

The U.S. Department of Agriculture (1970) has adopted standard rye grades (Table II) defined in the Official Grain Standards. Besides the limitations on heat-damaged kernels, U.S. grade specifications do not include any important

TABLE I
Canadian grades of rye

NO. 1. CANADA WESTERN—Minimum weight per measured bushel, 58 lb. Sound. Free from ergot after dockage removed. Maximum limits of foreign material including wheat: practically free.

NO. 2. CANADA WESTERN—Minimum weight per measured bushel, 56 lb. Sound. Practically free from ergot after dockage removed. Maximum limits of foreign material: matter other than cereal grains about 1/2%; cereal grains other than wheat: about 1-1/2%; total foreign material including wheat about 2%.

NO. 3. CANADA WESTERN—Minimum weight per measured bushel, 54 lb. Reasonably sound. Slightly damaged. Percentage of ergot after dockage removed: not more than 1/3 of 1%. Maximum limits of foreign material: matter other than cereal grains about 1%; cereal grains other than wheat, 3%; total foreign material including wheat: 5%.

NO. 4. CANADA WESTERN—Damaged with not more than about 5% heat damage. Percentage of ergot after dockage removed: not more than 1/3 of 1%. Maximum limits of foreign material: matter other than cereal grains about 2%; cereal grains other than wheat 7%; total foreign material including wheat: 10%.

CANADA WESTERN ERGOTRY RYE—Excluded from preceding grades on account of ergot. Percentage of ergot after dockage removed: over 1/3 of 1%. Maximum limits of foreign material: matter other than cereal grains about 2%; cereal grains other than wheat 7%; total foreign material including wheat: 10%.

CANADA WESTERN ERGOTRY RYE AND OTHER GRAINS—Excluded from preceding grades on account of admixture of other grains. Rye predominating. Percentage of ergot after dockage removed: over 1/3 of 1%. Maximum limits of foreign material: other than cereal grains about 2%.

NOTE—All grades of rye are required to be commercially free from seeds. No. 5 and No. 6 Buckwheat Sieves are used to remove seeds and broken rye.

quality factors. The following special grades have been defined:

Tough rye: rye with 14–16% moisture.
Smutty rye: rye having an odor of smut, or 14 or more smut balls in 250 g of rye.
Garlicky rye: rye containing two or more garlic bulblets in 1000 g of grain.
Weevily rye: rye containing live weevils or other insects.
Ergoty rye: rye containing more than 0.3% ergot.

In the U.S., only grades No. 1 and 2 are milled into flour for baking purposes. The lower grades are used in animal feeds.

EUROPEAN ECONOMIC COMMUNITY

The European Economic Community (E.E.C.) has defined standards for wheat, rye (Table III), barley, oats, and corn in Instruction No. 768/69 (Amtsblatt der Europaeischen Gemeinschaft, 1969).

In addition to the grading system, the E.E.C. has a guaranteed fixed-price system (intervention). Because of low market prices, rye is occasionally purchased at intervention prices. In 1975, rye for intervention had to meet the specifications given in Table IV.

Rye is classified as "bread rye" if the amylogram maximum is higher than 330 Brabender Units (BU). The rye standard of the E.E.C. does not include any

TABLE II
U. S. grades of rye

| | | Maximum Limits of | | | |
| | | Damaged kernels (rye and other grains) | | Foreign material | |
Grade No.	Minimum Test Weight per Bushel lb.	Total %	Heat-damaged %	Total %	Foreign matter other than wheat %
U.S. No. 1	56.0	2.0	0.1	3.0	1.0
U.S. No. 2	54.0	4.0	0.2	6.0	2.0
U.S. No. 3	52.0	7.0	0.5	10.0	4.0
U.S. No. 4	49.0	15.0	3.0	10.0	6.0

TABLE III
European Economic Community standard for milling rye

Moisture content	Max. 16%
Broken kernels	Max. 2%
Grain besatz (shrunken kernels, other grains, insect-damaged kernels)	Max. 1.5%
Sprouted kernels	Max. 1%
Black besatz (wheat seed, ergot, unsound grain, chaff, impurities)	Max. 0.5%
Hectoliter weight	Min. 71 kg

baking quality factors other than the limitation on sprouted and heat-damaged kernels. The identification "bread rye" implies an acceptable minimum baking quality.

GERMAN DEMOCRATIC REPUBLIC (G.D.R.)

The G.D.R. has adopted the rye standards given in Table V (Staatsverlag der DDR, 1973). Rye which is not within the specifications is classified as "feed rye."

OTHER COUNTRIES

Several other countries have defined rye standards or minimum values for baking strength. Sweden, for example, uses a continuous scale of the falling number values to calculate the price of rye. Poland is in the process of preparing specifications of standard rye grades (Ruebenbauer and Biskupski, 1953).

In contrast to wheat, the baking strength is not an important factor in rye grading. Some countries (for example, G.D.R. and West Germany) have started to include baking quality tests in their grading systems.

PERSPECTIVE

Consideration of compositional or biochemical factors in rye grading systems is a great problem because only simple and quick methods are suitable for commercial grading. So far, there has been no agreement among cereal chemists on the most suitable method for predicting baking quality. Nevertheless, a meaningful grading system of bread rye should include both internal and external factors to predict both milling and baking behavior. New methods should be developed for this purpose.

Seibel and Drews (1973) made a tentative proposal for classifying rye as "bread

TABLE IV
Maximum E.E.C. figures for intervention

Broken kernels	5%
Grain besatz	5%
Sprouted kernels	8%
Black besatz	3%

TABLE V
GDR standard for rye

Moisture content	Max. 18%
Black besatz	Max. 2%
Throughs—2-mm sieve	Max. 15%
Sprouted kernels	Max. 3%
Falling number	Min. 110 sec (9 g)

rye" based on three methods: maltose value, falling number, and amylograph curve (Table VI). These three tests are quick, simple, and precise, as required by commercial grading practices.

II. TESTS FOR MILLING AND BAKING QUALITY OF RYE

As indicated above, there are no quick and inexpensive tests for predicting the milling and baking qualities of rye. Many elaborate tests have come into use over many years. These will be considered in the sections that follow.

A. Tests Made on Grain (External Factors)

The following tests do not have any direct and significant relation to baking strength but are used in the rye grain trade.

MOISTURE CONTENT

The moisture content of rye during harvest may vary between 10 and 25%. Moisture content has an influence on the market value of the grain. Excessive moisture in grain promotes insect multiplication and self-heating, which can irreversibly damage the baking quality. Accordingly, it is understandable that moisture determination is one of the more important quality tests.

In the E.E.C., a standardized oven method is used for determining the moisture content (Amtsblatt der Europaeischen Gemeinschaft, 1969). Ground rye (minimum 90% through a 1-mm sieve) is dried for 2 hr in an air oven at 130° C. If the original moisture content is higher than 17%, the sample must be predried (whole kernels for 7-10 min at 130° C) before grinding. This method is also described in the International Association for Cereal Chemistry (I.C.C.) Standard No. 110 (International Association for Cereal Chemistry, 1960).

According to the Approved Methods of the American Association of Cereal Chemists (American Association of Cereal Chemists, 1969), a ground sample is dried for 60 min at 130° C if the moisture content is less than 18%. For samples with higher moisture contents, a two-stage air oven method is used.

Besides the air oven methods, chemical (Karl Fischer titration), electrical (dielectric meter method), and vacuum oven methods can be used to determine moisture content of rye.

For safe storage of bread rye, the following moisture contents have been proposed by Schaefer and Flechsig (1973):

TABLE VI
West German requirements of bread rye

Maltose value	Max. 3.5% dry basis
Falling number (7 g)	Min. 75 sec
Amylogram—90 g whole meal Gelatinization maximum	Min. 200 BU
Gelatinization temperature	Min. 63° C

Storage for several years:	below 14%
Storage for six to 12 months:	14–16%
Storage for a few weeks:	16–18%
Cannot be safely stored:	above 18%

BESATZ CONTENT

The German word "besatz" cannot be readily translated to other foreign languages. The term besatz applies to all components of a rye sample which differ from normal grain (Amtsblatt der Europaeischen Gemeinschaft, 1969). Table VII shows the classification of besatz.

The I.C.C. Standard No. 103 for determination of besatz is applicable to rye used for milling of bread flour (International Association for Cereal Chemistry, 1971a). It cannot be used for seed rye or feed rye. The most important besatz fractions relative to baking strength are "sprouted grains" and "unsound grains." The term "sprouted grains" applies to all grains in which the radical or plumule is clearly visible to the naked eye. "Unsound grain" covers all grains that have become unsuitable for human consumption because of rot, mold, bacterial attack, or other factors; it also includes heat-damaged grains. "Heat-damaged grains" are fully developed grains which show an endosperm of yellowish-brown to brownish-black color on dissection (Seibel, 1968).

The principle of besatz determination is to separate all the different besatz fractions from the normal grains by sieving or by hand picking. The separated fractions are weighed and expressed in percentage units.

DOCKAGE CONTENT

Dockage is not always the same as besatz. According to the official U.S. grain standards for rye, dockage includes weed seeds, weed stems, chaff, straw, grain other than rye, sand dirt, and other foreign material which can be removed readily from the rye by the use of appropriate sieves and cleaning devices, and undeveloped, shriveled, and small pieces of rye kernels which are removed in properly separating the foreign material and which cannot be recovered by rescreening or recleaning. The quantity of dockage is calculated in terms of percentage based on the total weight of the grain including the dockage. The percentage of dockage, so calculated, when equal to 1% or more, is stated in terms of whole per cent, and when less than 1%, is not stated. A fraction of a per

TABLE VII
Classification of besatz

Grain Besatz	Black Besatz
Broken grains	Wheat seeds
Shrunken grains	Ergot
Other grains	Unsound grains
Sprouted grains	Impurities
Insect-damaged grains	

Total besatz is the sum of grain besatz and black besatz.

cent is disregarded. The word dockage, together with the percentage thereof, shall be added to the grain designation (U.S. Department of Agriculture, 1970).

The main difference between besatz and dockage is that the dockage must be readily removable from the rye by sieves and other cleaning devices. The Carter Dockage Tester is used as a standard instrument. Normal dockage contents are 0–2%. On the other hand, besatz is removed by sieves and by hand, and includes rye grains which are unsound and unhealthy. Usual besatz contents are from 5 to 10%.

There is little or no relation between dockage content and baking quality of rye, because kernels that have been damaged by heating or sprouting are not included in the determination of dockage as required by the Canadian and U.S. rye grading systems. There is no difficulty in readily removing the dockage with simple cleaning devices prior to milling.

TEST WEIGHT

Test weight or weight per unit volume is used extensively as an index of grain quality. It depends on the intrinsic density of the grain and on density of packing of the grain in the measure used to determine the test value. The former factor reflects the chemical composition of the grain, whereas the latter depends more on kernel shape and size, presence of impurities, and degree of sprouting. Minimum test weight specifications are an integral part of the rye grading systems of Canada, the E.E.C., and the U.S. Over a wide range of test weights, the milling yield varies directly with test weight. With normal grain, there is no relation between test weight and baking quality.

In countries that use the Imperial system of weights and measures, the unit of test weight is the bushel weight in pounds per bushel. Canada and Great Britain use the Imperial bushel. The Winchester bushel measure is used in the U.S. Bushel weight is usually determined by weighing a pint of grain and multiplying the weight by 64.

In countries that use the metric system, the test weight is the hectoliter weight expressed in kilograms per hectoliter. It is determined by weighing 1 liter of grain on a Schopper Chondrometer. A special table is used to convert the liter weight to the hectoliter weight. Both the bushel weight and the hectoliter weight are expressed on an as-is moisture basis.

THOUSAND-KERNEL WEIGHT

The weight of 1000 kernels depends on grain density and size. Accordingly, it is considered a better measure of the physical condition of the grain than the test weight. To a limited extent, the 1000-kernel weight is directly related to the test weight. Traditionally, it has not been used as extensively as the test weight; however, now that the test has been automated it is being used quite widely.

Average 1000-kernel weights for rye grain of different origins are as follows:

Europe	above 25 g
Turkey, Australia	20–25 g
Argentina, Canada, U.S., U.S.S.R.	below 20 g

Rye with extremely low 1000-kernel weights gives low flour extractions. However, high weights (*e.g.*, tetraploid rye) do not always indicate a high flour extraction. There is no relation between 1000-kernel weight and baking quality.

SENSORY TESTS

Sensory evaluation of a grain sample is very important because taste and odor can sometimes detect deterioration in cereals that cannot be detected by analytical tests. Sensory evaluation results are now quite reproducible if properly trained testers are used.

Within the sensory test, it is necessary to differentiate odor and taste (Arbeits-gemeinschaft Getreideforschung, 1971). Regarding odor, the following assessments are made: normal, musty, sour, stuffy, and strange. For odor evaluation, the ground grain is mixed with cold water in a closed flask and placed in an oven at 45°C. After 20 min, the flask is opened and the odor is evaluated. Some odor types, *e.g.*, musty and stuffy, are quite deceptive, because they are also detected in freshly baked bread.

For taste evaluation, the following characteristics are used: normal, bitter, sour, sweet, and strange. In the actual test, ground grain is mixed with pure (taste-free) water. The suspension is poured into boiling water. The beaker is covered with a glass, and the tasting is carried out when the slurry has reached a temperature between 30° and 40°C.

Deviations in odor or taste can indicate detrimental biochemical changes within the grain. Rye for milling and bread-baking should have no abnormal odor or taste. Musty and stuffy odors are particularly dangerous to bread-baking, even if only 5 to 10% of affected rye is used in the milling blend. Most rye samples should be subjected to sensory evaluation because during growing, harvesting, and storage rye is not always treated as bread rye.

B. Assessment of Grain (Internal Factors)

COMPOSITION

The composition of the rye kernel is related to the shape and size of the seed which, in turn, depend on variety and on growing and ripening conditions. Climate, soil composition, fertilizer application, and agronomic practices influence the deposition of the constituents within the rye kernels.

The main constituents of rye grain are starch, proteins, pentosans, hemicellulose, cellulose, and mineral matter. Proteins, pentosans, hemicellulose, and cellulose are present in substantial amounts in the endosperm as well as in the bran, while starch is limited mainly to the endosperm. Shrunken kernels have an underdeveloped endosperm, a condition which can be detected by analysis for starch content. Rye grown in a hot climate with little or moderate rainfall produces small kernels which usually have a fully developed endosperm. Both types of kernels have a high proportion of bran and therefore contain considerable proportions of proteins, pentosans, mineral matter, and cellulose. In the case of shrunken kernels, the origin of these constituents is mainly the bran. In normal kernels, considerable amounts originate from the endosperm.

The composition of rye grain is strongly dependent on agronomic practices of the country or region where the rye is grown. Intensive cultivation practices in Germany and neighboring countries usually give high yields of grain but with reduced protein content and higher starch content. Such grain gives higher flour yields than grain of higher protein content grown under a different environment.

The percentage of ash of the whole grain is related to the proportions of endosperm and bran in the rye kernel. Environmental conditions combined with the intensity of cultivation can produce wide fluctuations in ash content. The extent of fluctuation in the composition of rye owing to varietal, environmental, and agronomic factors is illustrated by the data in Table VIII.

The differences in the content of the main constituents of the rye grain (Table VIII) that are particularly important technologically are those in mineral matter, protein, and pentosans. Variations in starch content are of lesser importance.

The main constituents of the rye kernel—starch, protein, and pentosans—exhibit special swelling properties when mixed with water. These properties make rye meal and flour suitable for the production of baked goods. Relative to the components of rye that swell on mixing with water, their solubility or swelling rate is of major importance. In this regard, it should be noted that rye contains matter which is two times more soluble in water than wheat. The higher solubility is mainly due to the presence in rye of higher proportions of pentosans, dextrins, sugars, and soluble protein. Because of the high pentosan content, rye proteins cannot form gluten. The amount of soluble substances is particularly important in relation to breadmaking properties.

A minor constituent that has special significance to baking quality is the enzyme content. Rye contains enzymes that attack all of its major constituents. The changes caused by the enzymes have marked effects on the solubility and swelling rates of the swelling substances, and thereby on baking quality. The enzymes that are especially important to baking quality will be discussed later.

Rye grain of different origins does not show marked variations in the amounts

TABLE VIII
Composition of rye grown in different countires
(All values on dry basis)

	1000-Kernel Weight g	Ash %	Protein %	Pentosans %	Starch %
U.S. (n=20)					
Average	18.9	1.70	13.8	8.3	60.2
Range	15.7–24.1	1.62–1.87	13.0–14.5	7.0–9.6	59.3–61.4
Canada No. 2 C.W. (n=12)					
Average	19.8	1.77	13.3	8.2	60.4
Range	18.2–22.1	1.61–1.86	12.4–14.0	7.6–8.8	59.0–61.3
Argentina (n=1)	12.0	2.16	15.0	10.0	56.0
U.S.S.R. (n=7)					
Average	17.9	1.74	13.4	8.1	60.2
Range	15.6–20.1	1.67–1.87	12.3–15.4	7.8–8.5	58.6–61.8
German varieties (3 years, three locations)					
Petkuser Normalstroh (n=9)					
Average	27.3	1.83	11.4	7.7	62.2
Range	19.0–33.7	1.67–2.16	9.0–13.4	6.6–9.0	59.3–64.1
Karlshulder Roggen (n=9)					
Average	25.2	1.97	12.6	8.5	60.2
Range	18.8–29.7	1.81–2.24	10.6–13.5	7.7–9.2	57.5–62.5

of soluble components. The solubility of the pentosans, however, can vary quite widely (Table IX).

The data in Table IX show that the region where the rye is grown has a definite effect on the amount of soluble pentosans and the proportion that it forms of the total pentosans. Rye grown in the wet climate of central Europe contains more soluble pentosans than rye grown in drier climates. This may be partly due to differences in enzyme activities caused by microorganisms developing on the grain during the wetter harvest conditions of central Europe. In addition to the preharvest effect of climate, the postharvest treatment (especially drying and storage conditions) of the grain can have a strong influence on the biochemistry of the grain. Because of highly variable climatic conditions in Europe, rye grown in this area varies quite widely with location. On the other hand, rye grown in North and South America is of more uniform composition and therefore of more uniform baking quality. Table X gives data on the composition of two varieties grown in two different years and at different locations in West Germany indicating the range of environmental effect within a relatively small climatic region.

ESTIMATION OF ENZYME ACTIVITIES

Since starch is the major constituent of rye flour, starch-degrading enzymes (amylases) play a key role in the baking quality of the flour. Rye flour can also

TABLE IX
Solubility of rye meal[a]

Origin of Rye	Soluble Material %	Soluble Pentosans	
		%	% of Total pentosans
U.S. (n=20)			
Average	14.6	1.40	17.0
Range	13.6–15.7	1.16–1.61	14.1–19.3
Canada No. 2 C.W. (n=12)			
Average	14.1	1.41	16.4
Range	13.5–14.5	1.35–1.52	14.9–18.5
Argentina (n=1)	16.4	1.75	17.5
U.S.R. (n=7)			
Average	14.3	1.39	17.8
Range	13.7–15.6	1.14–1.55	16.9–19.0
German varieties (3 years, three locations)			
Petkuser Normalstroh (n=9)			
Average	15.3	1.56	20.2
Range	13.7–17.0	1.21–1.87	16.3–23.3
Karlshulder (n=9)			
Average	16.0	1.80	21.3
Range	15.0–16.5	1.52–2.10	18.7–23.6

[a]Ground on Kamas Laboratory mill; values expressed on dry basis.

contain substantial levels of enzymes that attack proteins, pentosans, hemicelluloses, and celluloses present in the flour. In most flours, these enzymes contribute to the baking quality together with the amylases. However, the amylases are the most important group of enzymes in rye flour, relative to its baking quality. Accordingly, the estimation of amylase activity of rye flour will be discussed in some detail.

Amylograph Test. The first of a number of technological tests available for measuring amylase activity uses the Brabender Amylograph (Figure 1).

The amylograph is a torsion viscosimeter in which a stirred suspension of rye meal or flour is heated at a constant rate of 1.5°C/min, starting from 25° or 30°C. The viscosity of the suspension, through a predetermined temperature cycle, is recorded automatically as a curve (amylogram). The paper on which the amylogram is recorded has an arbitrary scale from 0 to 1000 Amylograph-Einheiten (AE) or Brabender Units (BU). In Europe, a simplified version of the amylograph, without cooling equipment, is used for assessing the loss of quality due to sprouting.

In the amylograph test, as the rye starch begins to gelatinize at about 55°C, the viscosity rises markedly. The peak height of the curve is reached between 62° and about 75°C. In the presence of starch-degrading enzymes, starch that is gelatinized by the rising temperature is rapidly hydrolyzed. This hydrolysis is reflected by a lower peak height (viscosity) and a lower temperature at which the peak occurs. The peak temperature is a better measure of amylase activity than the peak height. Peak temperature is also influenced by the gelatinization properties of the starch. Small starch granules, found in the outer layers of the endosperm near the bran, appear to be embedded in substances that inhibit their gelatinization. Higher temperatures are required to disrupt these substances. This produces an apparent increase in gelatinization temperature.

TABLE X

Composition of two varieties of rye grain
grown in different locations in West Germany
(Values on dry basis)

Variety Year	Region	1000-Kernel Weight g	Ash %	Protein (N × 6.25) %	Pentosan %	Soluble Pentosan % Total pentosan	Starch %
Nomaro	North	29.3	1.69	9.0	7.4	18.0	64.6
1972	Middle	29.8	1.72	10.1	6.6	23.3	63.5
	South	22.2	2.02	11.9	8.0	21.0	59.3
	North	19.0	2.16	12.8	9.0	22.4	59.5
1973	Middle	28.1	1.76	10.8	6.9	16.3	64.1
	South	29.0	1.90	13.4	8.2	17.6	61.2
Karlshulder	North	26.1	1.93	10.6	8.5	21.1	63.4
1972	Middle	27.3	1.81	11.5	7.7	23.6	61.4
	South	23.0	2.06	12.8	8.2	21.3	57.4
	North	18.8	2.24	13.5	9.2	23.4	58.4
1973	Middle	25.8	1.98	12.4	8.3	18.7	62.5
	South	29.4	1.86	15.6	8.8	18.9	57.5

Rye flours that have the same peak height but differ in peak temperatures are generally not comparable in baking quality. By way of example, the amylograph properties of flours from rye grown in the U.S. and Germany may be compared. U.S. rye flours often show a retarded gelatinization, yielding a peak of lower height but at relatively higher temperature. German rye flour, on the other hand, swells and gelatinizes at a high rate, and shows a relatively high peak viscosity at a lower peak temperature.

Peak viscosity in the amylograph is also influenced by the viscous properties of other swelling substances (pentosans and proteins) present in the flour. Meal particle size is another factor that affects peak height. For high test reproducibility, the rye grain must be ground to constant granulation. A variety of laboratory mills can be used for the grinding of grain for the amylograph test (Seibel and Drews, 1973).

It is not possible to identify the factors that affect the amylogram of a flour on the basis of the effects of single factors on the amylogram of a particular sample of starch. Because of extensive interactions, the total effect of a number of factors is not always additive. Some of the more important technological factors and their effect on peak height and peak temperature are given in Table XI. Figure 2 shows typical amylograms for the conditions given in Table XI.

Figure 1. Brabender Amylograph.

TABLE XI
Some factors affecting amylograms of rye meal

Meal	Amylogram	
Characteristic	Peak height	Peak temperature
1. Increasing α-amylase activity	Lowered strongly	Lowered strongly
2. Gelatinization rate		
A) High α-amylase	Raised	Lowered
B Low α-amylase	Lowered	Raised
3. Viscosity and water uptake of swelling substances		
A) High swelling substances α-Amylase (ICC units)		
a) Up to 1.0 units	Raised strongly	Raised strongly
b) Up to 10.0 units	Raised	Raised
c) More than 30.0 units	Raised	Raised or Unaffected
d) Over 400 units	Raised	Raised or Unaffected
B) Low swelling substances		
a) Up to 1.0 units	Lowered	Raised
b) Up to 10.0 units	Lowered	Unaffected
c) Over 30.0 units	Lowered	Lowered

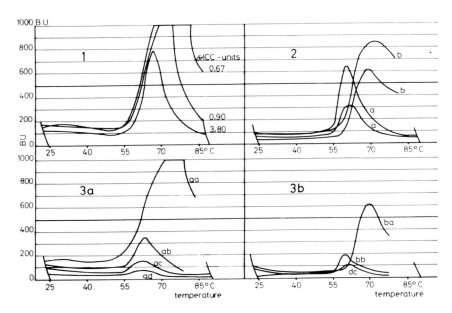

Figure 2. Amylograms of different rye meals (same grinder) under various conditions (see Table XI): 1) increasing α-amylase activity; 2) increasing gelatinization rate—a) high α-amylase samples; b) low α-amylase samples; 3a) high content of swelling substances with increasing α-amylase activity; and 3b) low content of swelling substances with increasing α-amylase activity.

In the standard amylograph test for rye meal, 90 g of ground grain at a moisture of 14% is used. For low α-amylase activities of up to 1.0 I.C.C. unit (International Association for Cereal Chemistry, 1971b), the viscosity in BU at 90°C can be used as the index of activity. Higher α-amylase activities can be expressed in terms of peak temperatures (Drews, 1973a).

Quick Amylograph Test. A new amylograph procedure has been developed (Brabender oHG, 1973) to decrease the time and the amount of meal needed for the test. The new procedure uses the same apparatus with a small bowl fitted inside the standard bowl (Figure 3). The standard bowl serves as a constant-temperature water bath. The test is made using 23 g of meal, 90 ml of water, and a bath temperature of 80°C. The peak of the amylogram is reached after about 4 min.

Falling Number Test. In this test for amylase activity, a suspension of meal, in a test tube of definite diameter, is stirred while heated in a boiling-water bath for 59 sec. The stirrer is then brought to its highest position and released. The viscosity of the gelatinized starch suspension decreases gradually due to enzymatic action. After a certain time, the viscosity will reach the point where it cannot support the stirrer. This time, plus the stirring time, is called the falling number. The falling number is inversely related to the activity of the starch-degrading enzymes. An apparatus that automatically records the falling number (Figure 4) is commercially available.

In contrast to the amylograph test, which gives a curve, the falling number test indicates the gelatinization behavior under the influence of diastatic enzymes by a number. Falling number correlates better with amylograph peak temperature than with peak viscosity. An exact relation between falling numbers and amylograph data should not be expected, since the time to reach peak height in the amylograph is much longer (up to 30 min) than the time in the falling number test (1–5 min). Furthermore, in the falling number test viscosity is measured at relatively high temperatures (90°–95°C), while in the amylograph the peak is registered at a lower temperature (60°–75°C).

Figure 3. Quick amylograph water jacket, bowl, and stirring paddle.

In the falling number test for rye, 7 g of rye meal is used. Samples that have falling number values below 70 sec should be tested, using 9 g of rye meal to increase precision (Drews, 1973a).

As with the amylograph test, falling number values are influenced by the gelatinization properties of the rye starch. At the same level of α-amylase activity, U.S. rye often gives higher falling number values than German rye. Differences in the water-binding capacity of the swelling substances of the rye meal can affect the falling number values. On the other hand, the α-amylase activity, which presumably is measured in the falling number test, depends on one factor only, the degree of sprouting. Deviations in falling number values due to secondary factors are relatively small for samples with low α-amylase activity (high falling number values), but they can reach considerable importance for samples of high α-amylase activity (low falling number values). Rye samples harvested from three locations in West Germany in 1973 gave three distinct relation between falling number value and α-amylase activity, depending on the location of growth (Figure 5).

Maltose Value Test. The maltose value, used as the index of amylase activity, is the amount of reducing sugars (expressed as maltose) formed after digesting a suspension of flour (or meal) for 1 hr at $27°C$ (Arbeitsgemeinschaft Getreideforschung, 1971). The amount of "maltose" after digestion is corrected for reducing materials (expressed as maltose) initially present in the flour. For routine quality control, this correction can be omitted to save time.

Maltose value is affected by the swelling substances (pentosans) found mainly in the aleurone layer of the rye kernel. U.S. rye is usually characterized by relatively low maltose values because its pentosans are less soluble (probably due to lower activity of the pentosan-degrading enzymes). On the other hand, German rye is characterized by relatively high maltose values; this is partly due to the higher activity of the pentosan-degrading enzymes. Golenkov *et al.* (1969)

Figure 4. Automatic falling number apparatus.

showed that in rye samples which were allowed to sprout under laboratory conditions, the degradation of pentosans occurs first, producing pentoses as reducing sugars. The formation of maltose occurs at a later stage. Accordingly, a high pentosanase activity produces a marked increase in the amount of reducing material and amylase activity.

For some rye samples, the maltose value is less accurate as an index of α-amylase activity than the amylograph or the falling number test. For rye samples with low α-amylase activity, the maltose value depends strongly on the viscous behavior of the aleurone pentosans.

Data for rye varieties grown in West Germany in a dry year (low α-amylase activity) gave significant correlations between amylograph peak height and maltose value, viscosity of soluble pentosans and maltose value, and amylograph peak height and viscosity of soluble pentosans (Drews, 1968; Seibel *et al.*, 1971; Table XII). For some rye samples, the three tests for α-amylase discussed so far

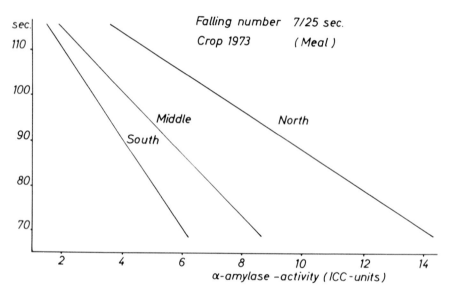

Figure 5. Relation between falling number value and α-amylase activity for grain grown in three regions of West Germany in 1973.

TABLE XII
Correlations for rye varieties of the
1967 crop of West Germany

	Correlation Coefficient
Amylogram peak BU:maltose figure (n = 76)	r = 0.91***
Viscosity of soluble pentosans:maltose value (n = 24)	r = 0.61**
Amylogram peak height BU:viscosity of soluble pentosans (n = 25)	r = 0.62**

give equivalent results. With such samples, any one test would be sufficient for quality control.

Dextrin Method for α-Amylase Activity. The dextrin method for measuring α-amylase activity was proposed by Lemmerzahl (1955). In this test, a special dextrin is added to the rye meal or flour suspension. After a relatively long reaction time (overnight) at room temperature, the suspension is filtered and the filtrate mixed with iodine solution. The solution shows the blue dextrin-iodine color if the α-amylase activity is low. If the activity is high, the solution is reddish or yellow in color. A color chart can be prepared for known α-amylase activities. The chart can then be used to determine the activity of unknown rye samples. This method for α-amylase activity is sufficiently simple for routine quality control.

Swelling Curve Test. In the amylograph and falling number tests, attention was focused on the swelling (gelatinization) of starch and how this phenomenon is affected by starch-degrading enzymes. In addition to starch, it is also necessary to consider the pentosans and pentosan-degrading enzymes as an integral part of the swelling behavior of rye flour when it is mixed with water.

Enzymes present in rye meal that degrade the pentosans, and thereby decrease their water-binding capacity, can be measured by recording the swelling curve with the amylograph. Such a curve can be obtained with the standard or the microbowl amylograph. In the actual test, a suspension of 120 g of meal and 410 ml of water is used. To promote enzymatic action, the reaction temperature should be about 42°C and the pH about 5.3 (Drews, 1971). Swelling curves obtained by this method for a number of different rye flours are shown in Figure 6.

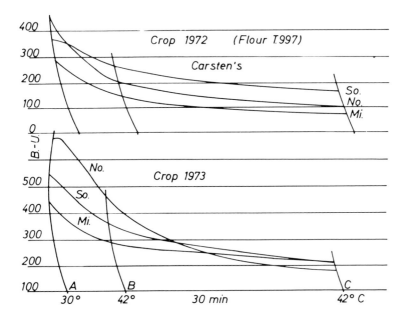

Figure 6. Swelling curves for rye from three regions in West Germany grown in 2 years.

The height of the starting point of the swelling curve, A, indicates the initial viscosity resulting from the water-binding capacities of the flour constituents immediately after mixing the suspension. In some cases, the swelling curve rises slightly after mixing. This is generally attributed to the time-dependent swelling that occurs in some flours. With continued mixing, the viscosity of the suspension usually decreases. The rate of decrease can be used as the index of enzymatic degradation of the soluble pentosans. The final viscosity (point C in the swelling curve) depends on the amount of swelling substances, the swelling or hydration properties of these substances, and the degree of enzymatic degradation during the test. The decrease in viscosity from point B to point C is also characteristic of the sample and a useful index of quality. Together with amylograph peak, gelatinization temperature, and falling number value, the swelling curve increases the possibility of predicting the baking quality of the rye flour much more accurately.

The swelling curve can also be obtained with the "quick" amylograph test. In this test, 40 g of meal or flour and 80 ml of water are used. The temperature of the water bath is held constant at 42°C and the curve needs to be recorded for only 10 min. A "quick" swelling curve for a typical German rye flour is shown in Figure 7.

DETERMINATION OF SPECIFIC ENZYME ACTIVITIES

α-Amylase. As mentioned above, the viscosimetric tests (amylograph, falling number) and the maltose value test do not always give an accurate measure of α-amylase activity. The values obtained by these tests are influenced by the differences in composition of the test material and by the solubility of the swelling substances in the flour or meal.

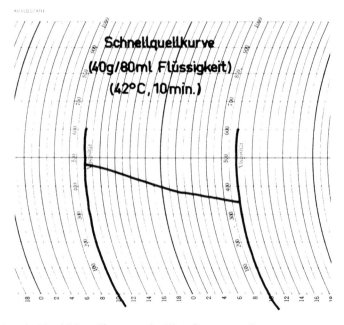

Figure 7. A typical "quick" swelling curve for West German rye flour.

A number of procedures for measuring α-amylase activity have been developed. The one applicable to rye meal and flour is the colorimetric procedure recommended by the I.C.C. (1971b). With this procedure, α-amylase activities in the range from 0.1 to 500 or more I.C.C. units can be measured. A critical evaluation of this method was carried out by Moettoenen (1969).

β-Amylase. β-Amylase is assumed to be present in rye in amounts sufficient for the degradation of damaged or soluble starch in the presence of α-amylase. There has not been much research on methods for determining the activity of β-amylase. The method published by Bendelow (1964) can be used for rye meal and flour.

Cellulase and Proteinase. Because of the highly complex composition and structure of rye cell-wall hemicelluloses, it is generally assumed that a number of specific enzymes are needed to degrade these substances. Although there is evidence that such enzymes are present in rye, they have not been purified and characterized. Accordingly, there are no procedures for measuring their activities.

The collective effect (or activity) of cellulases and proteinases can be estimated by the decrease of viscosity of a warm suspension of meal or flour (*e.g.*, swelling curve). Methods are also available for measuring the activity of enzymes involved in the loss of viscosity of cool meal or flour suspension. A method for estimating cellulase activity was developed by Rohrlich and Hitze (1970). Carboxymethyl cellulose is used as the substrate in this procedure. Proteolytic activity can be estimated using synthetic substrates by the procedure of Breyer and Hertel (1974).

Other Enzyme Systems. Water-soluble extracts of rye show more-or-less pronounced viscous properties that change with time because of enzymatic degradation of soluble pentosans (Holas *et al.*, 1973). Preece and MacDougall (1958) have shown that several specific enzymes are involved in the breakdown of these pentosans. The enzymes have not been isolated from rye.

IMPLICATIONS OF ENZYME ACTIVITIES ON BAKING QUALITY

It is not possible to consider the effect of a single enzyme in isolation on the baking quality of rye flour. The changes in processing behavior result from the composite effect of the many enzymes present in the flour. This effect is further complicated by variations in composition within the rye kernel and between samples of rye meal. If field conditions are such that enzymatic action occurs in the kernel (*e.g.*, sprouting), different parts of the kernel will be affected to a different extent. For example, α-amylase activity in sprouted grain is particularly high in the sub-aleurone layer. This layer is also rich in pentosans which surround the starch granules. This, in turn, protects the starch against amylolytic attack. On the other hand, if the pentosans are degraded, the same amylolytic activity will have a much greater effect on starch hydrolysis (as measured by the amylograph or the falling number test) than in the absence of degradation.

Enzymatic degradation of pentosans is particularly significant in rye grown in regions of moderately wet climate. Occasionally, under wet harvest conditions, microorganisms develop beneath the pericarp of the rye grain (Oxley, 1948). These microorganisms can contribute enzymes that degrade cellulose, hemicellulose, and pentosans. Through this action, the exogenous enzymes can have a marked effect on milling and baking qualities. Similar detrimental effects

TABLE XIII
Data related to starch and swelling substances
for Rye of different origin (Dry basis)

Origin	α-Amylase Activity ICC Units	Falling Number sec	Amylogram Peak		Maltose %	Soluble Pentosans %	Viscosity of Soluble Material sec	Water-Binding Capacity of Insoluble Material g/100 g
			Height BU	Temperature °C				
U.S. (n=20)								
Average	1.5			69	2.1	1.40	3.1	106
Range	0.24– 5.6	94–265	320–740	66–72	1.5–2.7	1.16–1.61	1.7–4.5	95–116
Canada No. 2 C.W. (n=12)								
Average	2.0			68	2.1	1.41	3.3	107
Range	0.98– 3.4	135–239	415–770	65–71	1.6–2.2	1.35–1.52	1.64–4.1	100–112
Argentina (n=1)								
Average	0.54	355	790	78	2.3	1.75	5.6	115
U.S.S.R. (n=7)								
Average	3.2			67	2.0	1.39	2.9	104
Range	0.39– 4.7	80–306	370–810	64–71	1.5–2.3	1.14–1.53	1.7–3.7	103–104
German varieties (3 years, three locations)								
Petkuser Normalstroh (n=9)								
Average	3.4			68	2.9	1.56	4.1	111
Range	0.34–15.0	72–320	250–820	63–75	2.4–4.0	1.21–1.87	2.4–6.6	92–124
Karlshulder (n=9)								
Average	2.8			70	3.1	1.80	5.4	121
Range	0.34– 9.5	99–336	420–>1000	65–79	2.6–4.0	1.52–1.80	3.3–7.6	112–136

of enzymes from microflora have been observed by Suomela and Ylimaeki (1970) in grain stored at relatively high moisture content. In addition to the direct effects of microfloral enzymes on quality, there is the indirect effect of increased α-amylase action owing to the release of the enzyme from kernel constituents or to an increase in the accessibility of its substrate.

Rye quality data related to carbohydrates and their enzymes are listed in Table XIII for flours milled from rye of different origin. The data indicate that a fairly wide range of values is obtained for any one test for samples from one origin. In general, rye grown in wetter climates (*e.g.*, in West Germany) has higher α-amylase activity and consequently lower falling number, amylograph peak viscosity, and amylograph peak temperature, and higher maltose values. Similar effects are indicated by the tests based on the pentosans and their degrading enzymes.

For one rye growing area, significant differences in test values related to rye carbohydrates can be obtained for the same type of rye grown in different years (Figure 8). In 1966, a very wet crop year in West Germany, the solubility of the rye pentosans was relatively high and the viscosity low. The maltose value was also higher than normal. The grain grown during a dry crop year (1967) had lower maltose values, higher viscosity, and lower content of soluble pentosans. It is obvious that, for a detailed evaluation of the rye carbohydrates relative to baking quality, it is necessary to use a number of different tests and to apply these to samples from a fairly small growing region. It is not satisfactory to make the tests on composite samples made up of grain grown in widely different climatic regions.

C. Assessment of Flour and Meal

FLOUR CLASSIFICATION

Rye flours can be classified on the basis of color or ash content. These two characteristics are directly related, especially for flours milled in one mill from similar grain (Brueckner, 1953). Accordingly, for routine quality control or commercial classification, either parameter can be used. Traditionally, ash content has been the main criterion for flour classification. Color values are used

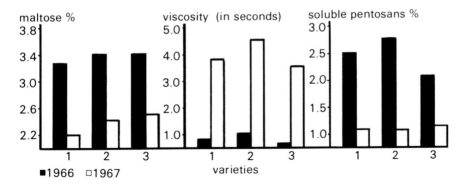

Figure 8. Variation in three quality parameters for three varieties grown in two crop years in West Germany.

to grade rye flours in some countries in eastern Europe (Zimmermann, 1968; 1969).

In West Germany, rye flours are classified according to ash content. Table XIV gives the 1975 ash values for the main classes of rye flour. The E.E.C. will also use ash content for rye flour classification; however, the number of different classes will be much lower than that shown in Table XIV.

In the U.S., rye flours are available in many different blends but there are no statutory standards. Three main categories of U.S. rye flour are distinguished, namely, white, medium, and dark. White rye flour is the patent grade. In ash content, it corresponds with the West German type 610 (Table XIV). Medium flour is the straight-grade flour; its ash content overlaps that of West German types 815 and 997. Dark rye flour, with an ash content of 1.0–2.0%, is a blend of the clear mill streams. According to Gordon (1970), a wide range of rye flours which are combinations of the three main categories is available to bakers.

Because of the soft character of rye endosperm, as compared with that of hard wheat, rye flours differ in texture from wheat flours. In particle size, rye flours are similar to wheat flours (Neumann and Pelshenke, 1954). The proportion of very fine particles is relatively high in rye flours but the weight fraction of these particles is small. The small proportion of very fine particles plays an important role in the baking properties of rye flours.

Rye flours that have a very fine and smooth texture, so-called dead-milled flours, should be avoided. Such flours contain too much soluble matter (starch, pentosans, and proteins). These substances swell easily when mixed into a dough, but the dough quickly loses its stiffness during fermentation. Such flours cannot be used for bread because the dough is too stiff (viscous) during the mixing stage and becomes too fluid for satisfactory dividing and panning. Bread baked from such flours will have a very weak crumb.

Particle size is also important in the case of whole rye meal used for bread production. For example, for pumpernickel bread the flour must be rigidly uniform in particle size for any one baking procedure. The appropriate particle size must be determined by trial. It is not possible to recommend a specific particle size for a particular type of bread.

Doose (1964) discussed the importance of flour particle fineness and the distribution of particle size of ground rye and rye meals in the production of dark rye breads. He distinguished rye meals of fine, medium, coarse, and extra-coarse

TABLE XIV
West German types of rye flour

Type of Flour	Ash Content (Dry basis)	
	min. %	max. %
610	0.58	0.65
815	0.79	0.87
997	0.95	1.07
1150	1.10	1.25
1370	1.30	1.45
1590	1.53	1.63
1740	1.64	1.84
1800 (meal)	1.65	2.00

granularity. A variety of milling procedures based on stones and corrugated rolls are used for producing rye meals. The meals produced on different mills may be similar in particle-size distribution but differ markedly in the nature of particle surfaces. Doose (1964) distinguished eight meals of different baking properties according to granularity and surface characteristics of the particles.

DETERIORATION DURING STORAGE

During normal storage conditions, rye flours deteriorate much faster than wheat flours. Whole rye flour cannot be stored safely as long as patent or straight-grade rye flours. Abnormal conditions such as high temperatures or high humidities accentuate the rate of deterioration during storage.

Storage deterioration of rye flour usually involves hydrolytic and oxidative changes in the flour lipids. This can usually be detected and measured by the so-called acidity tests. A suitable flour acidity test was developed by Schulerud (1934).

Flour acidity can also be measured by the standard fat acidity test. The fat acidity value is the amount (mg) of KOH required to neutralize the free fatty acids in 100 g flour. Fat acidity values of 40 or higher indicate an undesirably high level of deterioration.

Under some storage conditions, high flour acidity can result from microbial contamination. This type of deterioration usually involves other flour constituents in addition to lipids. It can be detected by measuring the pH of an aqueous flour suspension. The pH should not be lower than 5.5.

If there is any suspicion that a particular flour may not be sound, it should be subjected to detailed sensory evaluation. Freshly milled flour is not odorless and bland, but has a pleasant characteristic odor and taste. The development of an abnormal odor or taste during storage usually indicates that some deterioration has occurred. In some cases, the abnormal flavor disappears during the breadmaking process; in others, it is retained in the bread.

COMPOSITION

Chemical composition of rye flour can be a good indicator of its baking quality. Differences in composition arise from the fact that different parts of the rye kernel that are converted into flour vary widely in composition (Pelshenke, 1951; Brueckner and Schoenmann, 1961). The range of compositional variability can best be demonstrated by comparing the compositions of individual mill streams (Table XV).

In milling rye, it is relatively easy to obtain flours from the inner endosperm that will be relatively rich in starch and poor in swelling substances (pentosans and proteins). The break streams can be used to produce a flour of this type.

The outer parts of the endosperm yield middlings which are usually contaminated with various amounts of bran. Flours produced by the reduction of these middlings usually will be richer in pentosans and proteins than break flours. Reduction flours, obtained from the outer parts of the endosperm, are not only richer in pentosans but also contain a higher proportion of soluble pentosans. Accordingly, water uptake by the flour and viscosity of flour-water slurries generally increase with increasing extraction. This relationship holds for flours of ash content to about 1.1%. Amylograph and falling number data show parallel changes with composition that can be explained on the basis of total and

TABLE XV
Composition of rye mill streams (Dry basis)

	Break Streams						Reduction Streams				
	I	II	III	IV	V	VI	1	2	3	4	5
Yield (%)	18	10	14	7	6	4	12	3	5	3	3
Ash (%)	0.51	0.55	0.80	1.04	1.40	2.19	0.60	0.83	1.03	3.10	3.75
Starch (%)	82.9	81.5	75.4	68.5	66.5	55.4	80.7	73.6	69.3	43.3	32.5
Protein (N × 6.25) (%)	5.3	6.5	8.5	10.2	11.8	14.7	6.6	8.7	9.7	17.1	17.6
Pentosan (%)	3.0	3.4	4.2	5.7	5.3	8.1	3.3	4.7	5.6	10.3	14.0
Soluble pentosans (%)	1.3	1.4	1.8	2.4	2.0	2.3	1.4	2.0	2.3	2.5	2.6
Viscosity of water-solubles (sec)	14.5	15.3	17.1	24.4	17.4	19.5	15.3	18.5	22.3	16.7	17.4
Water absorption capacity of insoluble material g/100 g	90	99	113	150	131	156	97	120	132	158	175

soluble pentosan contents. Although the yield of flour in reduction streams is much lower than in the break streams (Table XIV), the effect of the reduction flours on the baking quality of a blended flour is nevertheless quite significant. However, straight-grade flours with ash contents to about 1.1% (*e.g.*, West German types 997 and 1150) usually have satisfactory baking quality. Dark flours are generally produced by longer milling flows; these show the expected effects of higher proportions of aleurone and bran constituents.

The compositions of the more important rye flours used in West Germany are given in Table XVI. These flours show variations in percentage of various constituents that can be related to baking quality.

IMPLICATIONS OF MILLING EXTRACTION ON BAKING QUALITY

The effects of extraction on baking quality can be examined in terms of the variations in the major constituents. It is also important to emphasize that the quality requirements of different flours (obtained by varying the extraction rate) will differ according to bread type, baking process, and regional preference. For example, in Canada and the U.S., light flours are used for the production of rye bread. In some cases, these flours are blended with various amounts of wheat flour. In West Germany, standard rye bread is made from 80%-extraction flour, while in the U.S.S.R., flour of 87% extraction is the major rye-bread flour. Other products (*e.g.*, pumpernickel, crisp bread, wafers, etc.) made in different countries are usually made from flours of slightly different quality (extraction). In the production of crisp bread (Knaecke), meal with falling number values as low as 80 is required in some countries, whereas others use flours with falling number values as high as 200. The protein content specifications are also usually quite different for the same baked product made in different countries.

The main effect of extraction on baking quality is through its effect on flour composition. This is discussed in the following paragraphs.

The most important constituent of rye flour is starch. It plays a major role in the crumb texture of baked bread. Starch is the substance that, in a partially gelatinized form, consolidates the crumb structure and determines its firmness. Therefore, a deficiency of starch or a poor starch quality can have a major effect on crumb characteristics. The changes in the breadmaking process that cause a release of water (*e.g.*, hydrolysis of pentosans, proteins, and starch), which must then be bound by the crumb constituents, have the same effects on crumb

TABLE XVI
Composition of West German flour types
(Dry basis, average values)

	Type of Flour				
	815	997	1150	1370	1740
Starch (%)	75.7	74.1	71.7	69.3	62.8
Protein (N × 6.25) (%)	9.6	10.1	10.6	11.2	12.4
Pentosan (%)	3.8	4.3	4.8	5.2	6.5
Soluble pentosans (%)	1.4	1.5	1.6	1.7	1.9
Water-solubles (%)	12.1	12.6	13.8	14.7	16.5
Viscosity of water-solubles (sec)	15.9	16.2	16.4	17.5	19.3
Water absorption capacity of insoluble material, g/100 g	113	117	123	133	148

TABLE XVII
Factors affecting crumb quality

Characteristic	Factor	Effect	Test
Diastatic state:			
a) Gelling properties of starch	Beginning and end gelatini-zation at low temperatures	Preharvest starch degradation	Amylogram peak height and peak temperature
b) α-Amylase	Activity over 1.0 ICC-units	Degrading the viscous properties of damaged and gelatinized starch	Peak temperature of amylogram; falling number; maltose value
Proteins	High protein content	Raise water absorption, when heat-denatured release of dough water	Farinogram; swelling curve
Pentosans Content	Low pentosan content	Low water absorption	Swelling curve; farinogram
Pentonase, etc.	Elevated activities of hemi-cellulases, pentosanases, proteases, etc.	Sprouting or microbial deterioration, lower viscosity of water-solubles, and higher solubility of swelling stuffs	farinogram; swelling curve
Degree of fineness	Influences the solubility of swelling substances and the amount of damaged starch	Promotes enzymatic action	Determination of the degree of fineness; maltose value
Buffering	Effects acidification by changing the pH value of the dough	Degree of extraction	Determination of ash content or color of flours
Dough yield	Excessive doughing water weakens dough consistency	Promotes enzymatic action	Measurements of dough viscosity

characteristics as a deficiency of starch. Table XVII lists some of the factors that influence crumb characteristics of rye bread; among these factors, starch and the enzymes that act on it are predominant.

If several of the factors shown in Table XVII are active simultaneously, the adverse effects on crumb structure will be much greater than expected. This will often be the case with flours milled from sprouted rye. On the other hand, it should be noted that only some of the same factors listed in Table XVII are involved in the firming of the crumb to an undesirable extent after baking. Starch and pentosan contents and their properties are especially important in this firming (staling).

Rye mill streams vary extensively in the proportion of starch and swelling substances (see Table XV). Accordingly, the swelling (dough-forming) properties of individual mill streams will differ widely. The effect of compositional differences resulting from varying extraction on baking quality is shown in Figure 9.

Figure 9 and Table XV show that good baking quality of mill streams can only be expected if the amount of starch compared with that of swelling substances does not differ from that found in mill streams with an ash content of about 0.9 (Drews and Reimers, 1965). The composition of flours obtained by straight or combined milling differs from that of mill streams of equal ash content. In low extraction flours, the main mill streams used will be those that are rich in starch. High extraction flours, on the other hand, include the streams that have high contents of swelling substances (pentosans and proteins). The influence of compositional differences of break streams and flours of increasing extraction on swelling curves is shown in Figure 10, which shows that viscosity of flour-water slurries increases with the extraction rate. Flour types from the same rye will often have similar falling number values and amylogram peak heights. Maltose values will usually be higher for the dark flour types because of the larger amounts of water-solubles in such flours. West German flour of type 1740 (high extraction) will often have a high falling number value but a low amylogram peak.

Figure 9. Rye bread from six consecutive break flours.

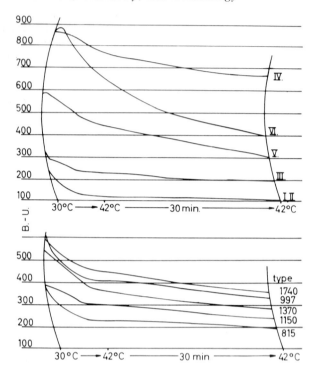

Figure 10. Swelling curves of rye break flours (top) and West German rye flour types (bottom).

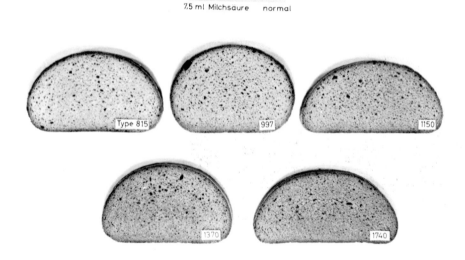

Figure 11. Comparative bread from West German rye flour types milled from the same grain.

The baking results on West German commercial flours of various extraction with European bread formula are shown in Figure 11. In general, baking results agree with compositional differences and results of quality tests. However, bread produced from flour type 1740 shows crumb characteristics which are not indicated by the falling number value. Stephan (1970a) suggested that the significance of falling number and other quality characteristics varies with extraction rate of the flour. Additional information concerning water absorption (obtained with the farinograph) is often useful for differentiating flours of different extraction (Weipert, 1972).

It is difficult to evaluate the contribution of gluten to the baking quality of rye and its interrelationship with extraction rate. Kosmina (cited by Golenkov, 1965) first demonstrated that rye gluten can be washed out, using the standard method of gluten washing from wheat flours, from the so-called interstitial protein obtained by first fractionating rye flour by specific weight. The content of interstitial protein in rye flour is relatively low, about 3.5%. Protein-rich products of rye milling, like the filter flour of pneumatic flour mills, can be used as the starting material for isolating interstitial protein. Rye gluten is of light gray color and absorbs water at a rate of two to three times its own weight. Its role in baking quality could be demonstrated using reconstituted flour made from rye gluten and rye starch. Pentosans added to interstitial protein affect its ability to form gluten. No gluten was obtained from the reconstituted flour when the proportions of protein and of pentosans resembled those of rye flour. Drews (1973b) showed that, with flours of ash content near 1.0%, factors related to the swelling substances influenced bread crumb firmness through their effects during the dough stage as much as did the starch and starch-degrading enzymes at the oven stage.

Table XVIII gives correlation coefficients between various compositional factors and dough consistency and the same factors and crumb firmness. These results underline the shortcomings of expressing baking quality by results from a single test.

Another quality factor of milled products that depends on milling extraction is the buffering capacity. Buffering capacity of the flour is important because it influences the degree of the acidification of the dough which is required to control enzymatic activity during the dough and early baking stages.

Buffering capacity of West German flours increased with extraction rate (Drews, 1972b). This was demonstrated by using a constant amount of lactic acid in the dough formula and determining dough or bread acidity by pH

TABLE XVIII
Correlation coefficients between dough consistency and crumb firmness and various flour (Type 997) properties

Property	Dough Consistency	Crumb Firmness
Protein content	−0.29	−0.18
Pentosan content	+0.68*	+0.24
Solubility of pentosans	−0.88***	−0.48
Viscosity of water-solubles	+0.73**	+0.89***
α-Amylase activity	−0.50	−0.90***
Falling number	+0.78**	+0.95***
Peak viscosity of amylogram	+0.80**	+0.93***
Swelling curve viscosity at endpoint	+0.86***	+0.74**

measurement. Flours that had a greater buffering capacity did not show the same drop in pH after addition of lactic acid (Table XIX).

Practical results of Doerner and Stephan (1956) using type 1150 flour indicated that the enzymatic degradation of starch during the oven phase was limited when the pH of the bread crumb was pH 4.3 or lower. Thomas and Luckow (1969) showed that pH was the limiting factor in the enzymatic degradation of starch in the oven phase by determining the starch content of dough and bread.

Drews (1970b) measured α-amylase activities of centrifuged flour-water slurries at different temperatures, pH, and salt concentrations. He showed that the three factors are involved in inactivating α-amylase as suggested previously by Huber (1964). At 40°C, α-amylase was inactive at pH 4.0 in the absence of

TABLE XIX
Rye bread pH and acidity

Flour type[a]	Lactic Acid 6.5 ml		Lactic Acid 8.5 ml	
	pH	Acidity[b]	pH	Acidity[b]
815	4.14	5.1– 6.0	3.94	6.1– 7.4
997	4.26	5.5– 7.1	4.08	6.3– 8.0
1150	4.40	6.2– 7.5	4.17	7.2– 8.7
1370	4.55	6.7– 8.1	4.27	7.5– 9.2
1740	4.62	8.0–10.1	4.39	9.1–10.7

[a]See Table XIV.
[b]Acidity = ml 0.1*N* NaOH.

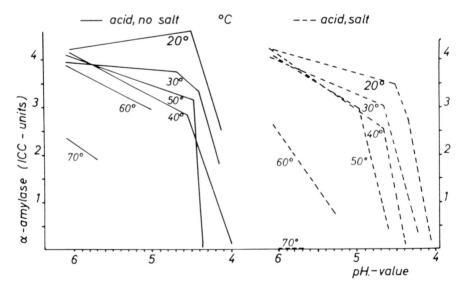

Figure 12. Influence of lactic acid and salt (1.5%, flour basis) on α-amylase activity of rye flour water-solubles at various temperatures.

salt; in the presence of salt the same effect was obtained at pH 4.4 (Figure 12).

The buffering capacity of rye flour is particularly important in the preparation of bread by the sourdough method. The pH of the crumb of sourdough bread increased significantly with increasing milling extraction as indicated by ash content (Figure 13).

With low extraction flours (ash content below 1.3%), acidification of the dough prepared by the sourdough method may be sufficient to inactivate α-amylase at the oven stage. In doughs from flour of higher extraction, the formation of fermentation acids increases at about the same rate, but a higher pH was maintained because of the higher buffering capacity of these flours. The higher amount of water-solubles (in high extraction flours), which usually increases further during the dough stage, also contributes to the buffering capacity. Furthermore, Rohrlich (1960) showed that phytin increased the buffering capacity of flour. Phytin content in flours generally increases with increasing milling extraction. Buffering capacity of flour is also affected by the particle size of the milled product. Finer flours give doughs with higher buffering capacity. This probably results from changes in solubility of flour constituents that are involved in the buffering action. Factors that affect the buffering capacity of rye-flour doughs are many. The contribution of each depends on the type of flour.

D. Quality Tests on Dough

WATER ABSORPTION AND DOUGH VISCOSITY

Water absorption of rye flour and dough viscosity (consistency) are extremely important properties relative to baking quality. The two properties are strongly interrelated.

Water absorption of rye flour is affected by many physical and compositional factors (Drews, 1966). It increases with increasing fineness of the flour. Flours with a high content of water-solubles will have high water absorption. The pentosans of the soluble fraction have a greater effect on water absorption than

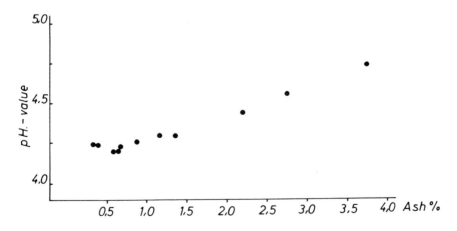

Figure 13. pH value of sourdough bread from rye mill streams of increasing ash content.

the proteins. In some flours, the soluble pentosan and protein contents are so high that it is not possible to use the flours for hearth bread because they do not have the strength to retain their shape during the proof or early oven stage. This behavior is accentuated by high α-amylase activity which tends to decrease the water-holding capacity of the dough.

Water absorption of rye doughs does not remain constant, but changes with time at a rate which depends on temperature, pH, and salt content. In the dough stage, the proportion of insoluble and soluble constituents is continuously changing due to enzymatic action. As the amounts of insoluble constituents are reduced, their contribution to water absorption rapidly diminishes. If enzymatic action is extensive, it will eventually destroy the water-binding properties of the water-soluble constituents. Analogous enzymatic factors are important in the production of wheaten bread but their detrimental effect on crumb firmness is minimal because wheat starch gelatinizes at a high rate only after most of the enzymes have been inactivated (Neukom et al., 1967; Lee and Ronalds, 1972) and because wheat flours contain less soluble pentosans (Bolling and Drews, 1970).

The viscous properties of rye doughs are extremely important to baking quality. This is in contrast to wheat doughs which must have an optimal balance between elastic and viscous properties for the best baking performance. Viscosity of rye doughs determines dough yield, stability, and volume, and bread loaf volume. Doughs of higher viscosity will yield more dough by retaining more water and will have better stability, but will yield lower loaf volumes.

The pentosans play the key role in rye dough viscosity. Proteins are important but not to the same extent as in wheat doughs. This is mainly due to the difference in the solubility of wheat and rye proteins after the flours are mixed into doughs. Only about 10% of wheat flour protein is soluble in water, whereas about 80% of

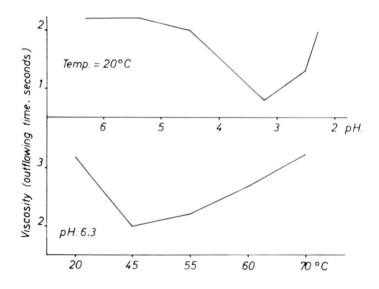

Figure 14. Viscosity of rye water-solubles at 20°C and different pH (top) and pH 6.3 at different temperatures (bottom).

the rye protein becomes soluble in the sourdough (Rohrlich and Hertel, 1966).

In relation to the viscosity of rye doughs, the viscous behavior of the solubles is extremely sensitive to variations in pH, temperature, and salt concentration (Figures 14, 15, and 16; Hagberg, 1952; Drews, 1969; Drews, 1970a).

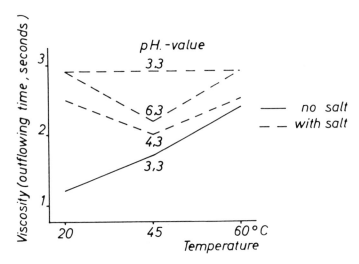

Figure 15. Influence of pH with and without salt on the viscosity of rye water-solubles; pH adjusted with lactic acid.

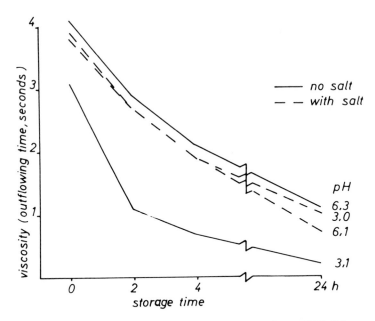

Figure 16. Change in viscosity of rye water-solubles with storage time at 20°C. Salt concentration was equivalent to 1.5% flour basis and pH was adjusted with lactic acid.

Under commercial baking conditions, dough viscosity can be controlled by manipulating dough temperature, pH, and salt concentration. The correct conditions will depend on the flour, the baking process, and the final product. The conditions cannot be predicted on the basis of any single quality test, but must be determined by trial.

Dough viscosity can change during processing. Vigorous mixing and other types of mechanical handling usually lead to a reduction of viscosity. Yeasted rye doughs will show a gradual drop in viscosity during fermentation; the rate of this drop depends on pH and salt concentration. In the sourdough process, there is an additional effect of various degrading enzymes produced by the acid-forming microorganisms. Under some conditions, the drop in viscosity can lead to doughs that are too fluid to process into bread. This can be partly overcome by using the salt-sourdough process of vom Stein (1971).

It is apparent from the foregoing discussion that there are many compositional, recipe, and processing variables that must be controlled to produce a dough with optimal physical properties for a specific breadmaking process.

E. Standard Baking Tests and Assessment of Bread Quality

BAKING TEST WITH YEAST

Commercial rye bread is generally produced with sourdough or a mixture of organic acids such as lactic and citric as the leavening agent. For scientific purposes, a baking test with yeast is commonly used.

The yeast test is based on the following recipe and processing conditions:

Flour	1000 g
Yeast	10 g
Salt	15 g
Water	as required
Dough temperature	29° C
Floor time	1 hr
Final proof	40–60 min
Bake time	60 min at 230° C

The amount of water is adjusted to give a dough of desired consistency as assessed subjectively by an experienced test baker. Any appropriate mechanical mixer can be used for mixing the dough. The dough is scaled and molded into desired shape. End of proof is judged by dough height.

The resulting bread is scored according to the procedure to be discussed later. This baking test is especially useful for assessing sprout damage.

BAKING TEST WITH LACTIC ACID

The lactic acid baking test is based on the following recipe and processing parameters:

Flour	1000 g
Lactic acid (80%)	6.5 or 8.0 ml
Yeast	10 g
Salt	15 g
Water	600–650 ml
Dough temperature	29° C

Floor time	60 min
Proof time	50–70 min
Bake time	60 min at 230° C

The effects of 6.5 and 8.0 ml lactic acid in the above recipe are equivalent to 40 and 50%, respectively, of sourdough. Water addition is adjusted to give a dough of desired consistency. Mixing can be by hand or with a mechanical mixer. Doughs are scaled and molded as required. Bread quality is usually assessed 15 to 20 hr after baking.

BAKING TEST WITH SOURDOUGH

The sourdough baking test corresponds to commercial rye bread baking. The sourdough used is prepared according to the so-called "Berliner-Kurzsauer-Fuhrung" method (Pelshenke, 1941; Schulz, 1947). Ninety grams of starter is mixed with 450 g flour and 405 ml water. This mixture is then allowed to ferment for 3 hr at 35° C to form the sourdough. The recipe and conditions for the sourdough baking test are as follows:

Sourdough	855 g
Flour	550 g
Yeast	10 g
Salt	15 g
Water	195 ml (approx.)
Dough temperature	29° C
Floor time	20 min
Final proof	45–65 min
Baking conditions	60 min at 220° C

As was the case in the two previous tests, the dough is scaled and molded into loaves of desired weight and shape. The 60-min baking time includes 5–10 min at the beginning at 250° C. Steam is usually injected at the beginning of the bake.

The three tests described above are standard methods used by the Detmold Institute for Cereals, Flour and Bread (Arbeitsgemeinschaft Getreideforschung, 1971).

AACC APPROVED METHOD

The approved method of the American Association of Cereal Chemists (1969) is based on a blended flour comprising 50% rye flour and 50% wheat flour of the clear grade. The procedure used is as follows:

1. Use one of the following formulas and fermentation schedules, depending on grade of rye flour to be tested.

	White and Light Rye Flour	Cream and Medium Rye Flour	Dark and Extra Dark Rye Flour
	Formulas		
	g	g	g
Clear flour	100	120	140
Rye flour	100	80	60
Yeast	4	4	4
Salt	4	4	4
Water	variable	variable	variable

	White and Light Rye Flour	Cream and Medium Rye Flour	Dark and Extra Dark Rye Flour
	Fermentation Schedules		
	min	min	min
First punch after	75	75	75
Second punch after additional	45	45	
Mold after additional	30	30	45

2. Place flour in mixing bowl, add 50 ml yeast suspension, 25 ml salt solution, and sufficient additional water to give desired consistency, and mix for 1 min.

3. Ferment dough according to appropriate schedule above and punch by any standard method. Mold by sheeting dough and rounding up in the hands. Place on transite block which has been sprinkled with fine corn meal. Proof to optimum height, give three light slashes across top of dough, and bake at 230° ± 5°C for 30–35 min. Use plenty of steam in oven, or brush loaf with water two or three times after crust has formed. Cool on a rack, weigh loaf, and measure its volume 1 hr after removal from oven. Place in fairly airtight cabinet until scored (preferably the following day).

SCORING OF RYE BREAD

Scoring of rye bread takes into account external characteristics (shape and volume) and internal characteristics (crumb texture and grain). Another

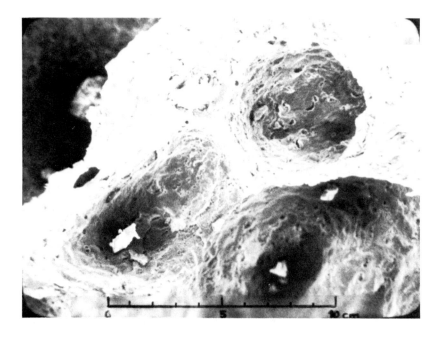

Figure 17. Photomicrograph of the crumb of rye-wheat bread (200×).

important aspect of rye bread scoring is sensory evaluation. In routine quality assessment, crumb grain is evaluated subjectively. For scientific purposes, the scanning electron microscope has been extremely useful for detecting differences in grain (Wassermann and Doerfner, 1974). Figs. 17 and 18 show the type of micrographs that can be obtained with the scanning electron microscope of crumb of rye-wheat mixed bread.

In West Germany a so-called 20-point bread score system has been used successfully for several years (Arbeitsgemeinschaft Getreideforschung, 1971). The factors used in this scoring system are as follows:

Shape, volume: The form can be flat, round, small, or unsymmetrical; volume can be measured by standard methods for this purpose.

Crust: Crust color can be light, dark, or uneven. The crust itself can be thin, thick, with cracks, or uneven.

Crumb: The evaluation of the crumb is divided into grain, elasticity, and texture. The grain is assessed visually; it can be closed, open, or uneven. The crumb may have cracks or a so-called water ring. Elasticity may be weak or unsatisfactory. Texture may be dry and rough, smooth, or wet and sticky. The crumb may ball during eating or be too weak for normal spreading of butter.

Sensory test: The taste can be sour, bitter, salty, musty, strange, etc.

The scoring system covers the four main factors enumerated above and is supplemented by the determination of the bread acidity. Normal acidity values are:

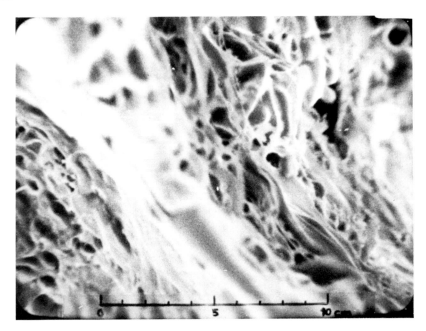

Figure 18. Photomicrograph of the crumb of rye-wheat bread (2000✕).

Rye-wheat mixed bread	7–9
Rye bread	8–10
Rye whole meal bread	8–14
Pumpernickel	higher than 14

Each of the four factors and acidity values are assessed and given a score. The score is the sum of individual scores. Rye bread without any internal and external defects and with normal acidity will have a total score of 20 points. The maximum scores for the factors described above are given in Table XX.

Based on the 20-point scoring system, bread quality can be expressed as:

Quality	Total Score
Excellent	19–20 points
Good	17–18 points
Satisfactory	16 points
Unsatisfactory	lower than 16 points

III. RYE AND MIXED RYE-WHEAT BREAD

A. Commercial Production

DOUGH-MIXING PROCESS

In the production of rye bread, the dough-mixing process is different from that used for wheat flours. Rye doughs are mixed slowly, without great energy input. If the energy input is excessive, rye doughs become too tough or bucky because of the development of very viscous properties of the pentosans. This is undesirable. There is an optimum mixing time for each type of dough depending on the composition and enzymatic activities of the flour.

In the production of mixed rye-wheat bread, dough development can include more intensive mixing and even high speed kneading, especially when the wheat flour predominates in the flour mixture. Wheat flours, in contrast to rye flours, have a much lower content of both insoluble and soluble pentosans. As the proportion of wheat flour increases, the influence of the pentosans on the dough properties is gradually diminished. At the same time, a higher energy input is necessary to develop the gluten. As the effects of the pentosans diminish, according to the lower amount of rye flour in the mix, the gluten-forming properties of wheat flour become more important.

TABLE XX

Detmold rye bread scoring system

	Maximum Score, Points
1. Shape, volume	2
2. Crust	2
3. Crumb	
a) Grain	5
b) Elasticity	3
c) Texture	2
4. Sensory test	5
5. Acidity	1
Total	20

WATER ABSORPTION, DOUGH YIELD, AND DOUGH STRUCTURE

As discussed earlier, water absorption, dough yield, and dough structure depend on the flour type used for dough preparation. Rye flours of high extraction have a high water absorption and produce high dough yields, but the doughs tend to be tough and sticky. This behavior is due to the high amounts of water-solubles in such flours. Especially sticky doughs will lead to extremely low dough volumes.

Doughs prepared from rye-wheat flour mixtures give zymotachygrams that resemble those for wheat doughs. This indicates that gluten formation of the wheat flour is quite feasible in the presence of low amounts of soluble pentosans from the rye. Addition of small amounts of rye flour to wheat bread can have a beneficial effect (Stephan, 1963). Casier *et al.* (1973) showed that improvements in water absorption, volume, and keeping quality (freshness) of baked wheat goods can be obtained by adding purified water-insoluble rye pentosans. According to Jankiewicz (1975), addition of rye pentosans to strong wheat flours diminishes the solubility of the wheat proteins.

Dough yields for a series of West German flours of increasing ash content for two types of dough are given in Table XXI. In each case, dough consistency was adjusted to commercial practice. For both dough-making processes, the dough yield increased with increasing ash content (extraction) of the flour.

Dough yields for two reciprocal rye-wheat flour blends are given in Table XXII. The rye flour was type 1150 and the wheat flour was type 1050 (ash, 1.05%; protein, 13.5%). The 60:40 rye-wheat flour blend gave a higher dough yield at the three levels of consistency than the 60:40 wheat-rye blend. Rye flour contributes more to dough yield than wheat flour.

Wheat starch binds more water in dough during gelatinization than rye starch; accordingly, baking results with flour mixtures in which wheat flour predominates are improved by using higher water absorptions. However, a

TABLE XXI
Flour type and dough yield

Flour Type[a]	Dough Behavior	Dough Yield/100 g Flour	
		Sourdough process	Straight-dough process
815	Normal	163	168
997	Normal	167	172
1150	Normal	169	174

[a]See Table XIV.

TABLE XXII
Flour mixture and dough yield

Dough Behavior	Dough Yield, g/100 g Flour	
	70% Wheat	70% Rye
Firm	156	160
Normal	162	166
Soft	168	172

firmer dough consistency is necessary in producing mixed rye-wheat bread when the flours are rich in α-amylase or when the wheat flour is particularly high in protein content. These conclusions have been confirmed by the work of Bruemmer and Stephan (1972b).

FERMENTATION

In the straight-dough processes, yeast is used as the leavening agent and the dough is acidified artificially by using organic acids such as lactic acid or citrates of acid character. This yeast fermentation procedure is more dependent on the viscous behavior of the swelling substances of rye than the sourdough process. In some cases, the swelling substances are so viscous that it becomes necessary to lower dough consistency by adding hemicellulases before fermentation (Rotsch and Stephan, 1966). In such cases, the keeping quality of the bread may be improved simultaneously (Drews, 1972). In the sourdough process, there is a strong enzymatic degradation of the pentosans by enzymes from the sour microorganisms, so that it is not necessary to use hemicellulase supplements (Rehfeld and Kraus, 1961).

The sourdough method is a sponge-and-dough process. It includes a starter, which already contains acid-forming bacteria and sourdough yeasts. The sourdough is the leavening agent. The starter is activated by adding flour and water and allowing the mixture to ferment. In the perpetuated or multiple-stage process, this starter is first increased to the basic sour, then to the full sour which, after a fermentation period, is used to prepare the dough with addition of the remaining flour, water, salt, and special ingredients.

The biochemistry of sourdough fermentation was reviewed in detail by Rohrlich (1961) and Schulz (1966). The main acids formed in the sourdough process are lactic and acetic. *Lactobacillus plantarum* produces lactic acid only, while *L. brevis* produces both acetic and lactic acids. Accordingly, it is possible to change, to some degree, the ratio of the two fermentation acids by favoring the development of either species by varying the dough conditions.

In modern sourdough bread technology, the entire process has been simplified and automated. The multiple-stage procedure has been replaced by shorter methods based on one or two fermentation steps. A widely used process in Europe is the Berlin short sour process (Pelshenke, 1941; Schulz, 1947). Another process, known as the Detmold sourdough method, has been developed (Stephan, 1960; Stephan, 1970c). In this process, leavening is augmented by the addition of baker's yeast. The scheme for the two-stage Detmold process based on 100 kg flour (70% rye flour, 30% wheat flour) is shown in Table XXIII. In this scheme, the first stage gives 25 kg basic sour (1 kg starter, 16 kg flour, 8 kg water). One kilogram of this sour is removed to be used later as the new starter for the next batch of dough. Twenty-four kilograms of the basic sour is used in the second stage. In preparing the dough, the remaining 60 kg of the flour is added to 72 kg full sour as well as 30 l. of water, 1.3 kg yeast, and 1.7 kg of salt. The dough is mixed mechanically. Final dough temperature is about 27° C. After resting for 10 min, the dough is cut, molded, and proofed as single dough pieces.

In modern sourdough bread production, it is possible to automate the procedure (Schulz and Stephan, 1958) and to store the sourdough at reduced temperatures for a certain time without diminishing its acid-forming properties (Spicher and Stephan, 1969).

Another modification of the one-step method of sourdough production is the salt-sour process (vom Stein, 1971). Since salt affects the metabolism of microorganisms (acid-forming bacteria or yeast), sourdough production is carried out at higher temperatures in the presence of salt (starting temperature 30°–35° C) and longer fermentation times (18–24 hr). The process is easy to control and can be used for the production of good quality bread. This process usually gives a slightly higher yield of dough and bread than the standard sourdough process because of smaller fermentation losses. However, it requires higher amounts of baker's yeast. In cases where it is desirable to maintain the aroma of traditional sourdoughs it is possible to acidify the dough slightly by reducing the amount of sourdough and adding some organic acids (combined method).

In the production of mixed rye-wheat bread, most of the rye flour is generally used for making the sour. Wheat flour is added to the sour to make the final dough. Wolter (1974) has developed a sourdough procedure that uses a rye-wheat flour blend in the sour.

INFLUENCE OF BAKING INGREDIENTS

The quality characteristics of flour that affect rye or mixed rye-wheat bread quality have already been discussed. Of the minor baking ingredients, salt has a definite effect through its influence on the solubility of flour proteins and on the activity of the technologically important enzymes. The quality of the water can influence the physical properties of dough or the fermentation process. Certain salts present in water (*e.g.*, calcium salts) may affect the viscous behavior of soluble swelling substances or may affect the activity of starch-degrading enzymes. These effects are normally negligible since the pH of rye doughs is usually quite low.

Dough improvers are not used extensively in rye and mixed rye-wheat bread production. Some of the commonly used improvers are pregelatinized potato flour or pregelatinized corn or rice starch added at about 3%. These additives have a number of beneficial effects in rye doughs and bread. Because of the high water-binding capacity of pregelatinized flours (or starch), they contribute to the water uptake (absorption) of the doughs. Subsequently, they increase dough and bread yields. If the rye flour is deficient in swelling substances, addition of pregelatinized flour improves dough consistency. Pregelatinized flours can contribute to the firmness of the bread crumb, especially when the basic rye flour is milled from sprouted grain. Acidified pregelatinized flours are now

TABLE XXIII
Dough-making scheme of the Detmold two-stage process

Starter for Basic Sour kg	Dough Yield %	Absorption %	Temperature ° C	Ripening Time hr	Flour kg	Water l.
Stage I: Basic Sour						
1	150	50	22–26	15–24	16	8
Stage II: Full Sour						
Basic sour 24	180	80	27–33	2.5–3.5	24	24

commercially available which can be used directly in the straight-dough process of rye bread production.

There are a number of other substances that are used in rye bread production to regulate absorption or to improve dough-handling properties. The most common of these are hydrocolloids or polysaccharide gums such as locust bean or guar gum. The effects of these materials in the dough resemble that of pregelatinized flours. As these gums are free of starch, their effect in the oven phase differs from that of the pregelatinized flour. They extend the shelf-life of the bread.

Other additives used in rye bread production are of enzymatic nature. Since rye flours usually contain sufficient α-amylase, bacterial α-amylase is used only occasionally for its specific beneficial effect (Schulz and Uhlig, 1972). Specific effects of proteases and hemicellulases or cellulases in rye doughs have already been mentioned. Flours lacking these enzymes need supplementation to ensure sufficient gelatinization of the starch in the oven phase required for satisfactory crumb.

FERMENTATION TIME, BAKING TEMPERATURE, AND BAKING TIME

Fermentation time depends on the composition of the flour, the activity of the flour enzymes, dough recipe, and fermentation temperatures. The baker, producing large amounts of bread, must also recognize the effects and limitations of his processing equipment. In addition to selection of appropriate flour, the added ingredients must be of the right specifications to ensure normal running of the production process and good bread quality.

Bread quality is influenced by the baking process; baking temperature, uniformity of temperature in the oven, and baking time are all important (Thomas and Tunger, 1965). Optimum temperature and time depend on the weight of the dough pieces and the shape of the loaves, both of which determine the rate of heat transfer to the inner parts of the loaf. The rate of heat transfer affects a large number of interrelated biochemical, chemical, and physical processes which combine to give a satisfactory loaf.

Many rye bread bakeries still use steam in the oven, especially at the beginning of the baking stage. Appropriate use of steam is important for the development of desirable crust characteristics.

Table XXIV summarizes the baking conditions generally used in West Germany for the production of 1.5 kg hearth loaves. Pan rye bread is usually

TABLE XXIV
West German commercial baking conditions

| Bread Type | Baking | |
	Temperature °C	Time min
Rye bread	320 → 230[a]	60
70% Rye–30% wheat	305 → 220[a]	60
30% Rye–70% wheat	240 → 220[a]	60

[a]Descending baking temperatures: short prebaking at high temperature; after transposing the loaves, final baking at lower temperature.

baked at lower temperatures but for longer times.

The baking process is very important in rye bread production relative to bread quality. It is during this phase that suitable crumb characteristics are developed. The actual procedure by which the optimum crumb structure is produced depends on the type of bread being produced and the oven used. A particularly important parameter during the baking phase is the crust-to-crumb ratio (size of loaf).

The baking phase is also important in relation to bread yield. With standard West German bread, baking conditions give a baking loss of 13% in the production of rye or mixed bread and 13.5% when wheat flour prevails in the flour mixture in the production of 1.5-kg loaves. Average bread yields are 135–138 kg of bread from 100 kg of flour, and depend on the percentage of rye flour.

Baking conditions affect a number of characteristics of the finished bread, for example, loaf volume and flavor. Aromatic properties of rye bread are related to the thickness of the crust formed in the baking process. The special flavor produced by sourdough can be significantly altered by baking conditions. Quite frequently, the baker's success depends on the performance of his oven.

B. Shelf-Life and Mold Prevention

Bread staling is of great economic significance. Staling of wheat bread has been extensively investigated in many research laboratories; much less has been done on staling of rye bread. In relation to staling of wheat bread, attempts have been made to retard the staling process by adding enzymes, emulsifiers, water-binding substances, and fat. By using optimum dough and recipe conditions, it is now possible to increase the shelf-life of wheat bread by several days by the use of permitted additives.

It has been amply demonstrated that retrogradation of the starch is the main cause of bread staling. Banecki (1970; 1972) has shown that proteins are also involved in the physical changes that occur during staling. He noted that the viscosity of extracted rye protein changed quite markedly during the first 48 hr after baking.

Many other changes in bread properties are considered to be part of the staling process. The crust changes from crisp and brittle to soft and leathery while the crumb changes from a soft to a firm texture. There is also a gradual, but quite noticeable, alteration of flavor. The staling problem with rye bread is far less serious than with wheat bread. Whole-meal rye bread can be kept for at least 4 weeks (if properly wrapped). There are no significant changes in rye and mixed rye-wheat bread within 1 week if the bread does not dry out.

RETARDATION OF STALING

Shelf-life of commercial rye bread can be extended by the following processing or ingredient factors:

—use of sourdough
—optimization of the enzyme activity of the flour
—addition of malt flour
—use of high baking absorption
—addition of pregelatinized potato flour or starch (3%)

—addition of dried and milled old bread (Restbrot, 3%)

—prebaking for 1–2 min at 400°–500° C

—wrapping or packaging while still warm (80° C)

Addition of crumb softening agents or moisture retaining substances has little practical value in rye bread production. Bruemmer and Stephan (1972a) showed that the beneficial effect of special crumb softeners in mixed rye-wheat bread diminishes as the proportion of rye flour is increased in the recipe.

MOLD PREVENTION

Mold growth does not occur very often in rye bread made by the sourdough process. Infections can result from direct contact with mold spores or from airborne spores (Spicher, 1970). It has been shown (Arbeitsgemeinschaft Getreideforschung, 1973) that the use of sourdough prevents mold growth for several additional days compared with bread produced by a recipe using acids like lactic, citric, or tartaric. Addition of calcium propionate (0.4%) gives some protection against mold growth, especially if the bread is sold in sliced form. Sorbic acid (0.2%) is another approved anti-mold agent. Mold growth can be inhibited without propionic and sorbic acid by increasing baking time. Extension of baking times will increase baking losses. Absolute protection against mold can only be achieved by sterilization of the wrapped bread with saturated steam or heated air. A temperature of 90° C inside the bread for at least half an hour is required for complete sterilization. In practice, this can be achieved by treatment for 150 min with air at a temperature of 110° C (Schulz, 1960). In commercial practice, mold development is not nearly as serious a problem in rye bread as in wheat bread.

C. Packaging and Storage

Most of the rye bread produced around the world is sold in unwrapped form. However, wrapping is becoming more popular for a variety of reasons. Wrapping protects the bread against moisture loss and mold spore infection.

The packaging material should be transparent to show the crust and the crumb, if portions of sliced loaves are sold.

To retain bread quality, and especially the flavor, it is necessary to select appropriate packaging material. The material can be tested as follows: Wrap a warm loaf (about 30 to 60 min after baking) in the packaging material and store for 24 hr at 20° C and 60% relative humidity. If the loss in weight is less than 0.5%, the packaging material is satisfactory for maintaining crumb softness and bread flavor (Stephan, 1970b). The wrapping material is unsatisfactory if the weight loss is substantially higher than 1%. If the loss in weight is higher than 3%, staling will not be retarded and the crust will become very hard as in unwrapped bread.

For packaging rye bread and mixed rye-wheat bread, polyvinylchloride film is excellent. For whole-meal rye bread, good results have been obtained with aluminum foil wrapping. In addition to retaining the quality of the bread, good packaging material increases the shelf-life from days to weeks.

D. Nutritional Value

Cereal grain products form a high percentage of our daily diet. In West

Germany, for example, cereals provide 21% of the calories, 38% of the carbohydrates, and 20% of the protein in the human diet. Only meat products contribute more protein to the diet.

Different types of bread vary significantly in caloric content (Table XXV) (Cremer *et al.*, 1969; Steller *et al.*, 1974). The caloric content of rye bread is always lower than that of wheat bread.

In overall nutritional value, rye bread has some advantages over wheat bread (Seibel, 1975; Table XXVI). It is lower in caloric content, and higher in mineral and fiber content. Rye protein is somewhat better in nutritional quality because of its higher lysine content (see Section IV).

Significant variations in nutritional factors can also be readily demonstrated for rye flours of different extraction (Thomas, 1964; Table XXVII). Rye flours of high extraction contain substantial amounts of calcium, phosphorus, iron, and vitamin B_1. In countries where rye bread and mixed rye-wheat bread form a significant portion of the daily diet, it is not necessary to enrich wheat flour with calcium, iron, and vitamin B_1.

IV. RYE BREAD AND BAKED PRODUCTS AROUND THE WORLD

A. North America

In North America, rye bread is produced in many different forms to meet the

TABLE XXV
Caloric content of different bread types

Bread Type	Weight per Slice (g, Dry basis)	Calories per 100 g
Whole meal rye	29	210
Rye	25	230
Rye-wheat mixed (50:50)	22	250
Wheat toast	16	275
Crisp bread	9.5	350–390

TABLE XXVI
*Differences between wheat and rye bread
relative to nutritive value*

	Wheat	Rye
Dough yield, g/100 g flour	153–155	166–168
Volume, cc/100 g flour	600	300
Bread yield, g/100 g flour	130	145
Caloric content	+[a]	−[b]
Protein content	+	−
Protein efficiency	−	+
Mineral content	−	++[c]
Fiber content	−	+
Vitamin content	+ −[d]	+ −

[a]+ = High.
[b]− = Low.
[c]++ = Very high.
[d]+ − = Normal.

demands of consumers. There are both round and elongated loaves, baked with or without pans. North American rye bread recipes use a variety of minor ingredients in addition to the basic components of flour, yeast, salt, and water. The following minor ingredients are used in certain rye breads:

—yeast foods as a fermentation stimulant

—sugar and malt to improve crust color and fermentation

—shortening to improve keeping quality

—molasses to improve taste and increase darkening of crumb and crust.

Four basic types of rye bread are produced in the U.S. (Weberpals, 1950).

AMERICAN RYE BREAD

This is a light rye bread with good grain and soft texture. It is usually made from a blend of 60–85% first or second clear wheat flour and 15–40% rye flour. Yeast, salt, and water are the other ingredients. Sourdough is not used in the production of this type of bread.

SOUR RYE BREAD

This is a heavy rye bread. It is made by the sourdough process. The sour is made from old dough, rye flour, and water. First or second clear wheat flour is used at the dough stage, together with salt and water as required.

PUMPERNICKEL

Both light and dark pumpernickel bread are produced in the U.S. The U.S. pumpernickel bread is quite different from the original German pumpernickel. It is made by the sourdough process. The sour is made from old dough, rye flour, and water. Flour, added at the dough stage, comprises 80–90% wheat flour and 10–20% rye meal. Salt and water are added as required. Other ingredients include shortening (2%) and molasses (2–4%). Lower grades of wheat and rye flour are used to produce the dark pumpernickel.

SWEET OR PAN RYE

This is a light, sweet rye bread with excellent flavor. It is made by the straight-dough process. The formula includes 10–40% white rye flour and 60–90% patent or clear wheat flour. In addition to yeast, salt, and water, the formula includes shortening and syrup.

TABLE XXVII
Composition of rye flours of different extraction
(12.5% Moisture basis)

	Extraction Rate					
	45	60	70	82	87	98
Protein (N × 6.25), %	5.8	5.9	6.4	7.6	8.4	8.6
Fat, %	0.6	1.1	1.2	1.6	1.8	2.2
Carbohydrates, %	80.2	79.4	78.6	76.3	75.1	73.4
Fiber, %	0.34	0.61	0.67	0.88	1.01	1.71
Ash, %	0.47	0.57	0.74	1.16	1.30	1.62
Calcium, mg %	20.8	21.8	20.9	25.3	28.1	38.7
Phosphorus, mg %	85	95	133	214	249	410
Iron, mg %	1.91	2.08	2.42	2.75	2.86	3.56
Vitamin B_1, γ/100 g	165	212	263	314	357	403

B. Europe

STANDARD RYE BREADS

The most important rye-consuming countries in Europe are Austria, Czechoslovakia, Poland, East and West Germany, the Scandinavian countries, and the U.S.S.R. The following bread types are produced:

—rye bread: produced from 100% rye flour (ash content of 0.8–1.6%)

—rye-wheat mixed bread: produced from a mixture of rye and wheat flours containing at least 50% rye flour

—wheat-rye mixed bread: produced from a mixture of flours containing at least 50% wheat flour and at least 10% rye flour

—whole meal rye bread: produced from whole meal rye flour

Besides the basic types of rye bread noted above, there are breads produced from various mixtures of whole meal rye flour, whole meal wheat flour, rye flour, and wheat flour.

CRISP BREAD (KNAECKEBROT)

Beginning in 1900 in Scandinavia, crisp bread developed into a staple bread. Crisp bread has excellent storage properties. It is still used extensively by forest workers in northern Scandinavia. It is light to carry and can be eaten immediately after storage at very low temperatures. Today, crisp bread is sold in many countries as a specialty product.

Crisp bread is produced mostly from whole rye meal with or without small amounts of flour. One type of crisp bread is brown and the other is white. In mixing the dough for this type of rye bread, either milk or water is used. Brown crisp bread is produced from yeast-leavened dough. The lighter types do not use yeast. The dough is kept cool. It retains the air mixed into it throughout the entire dough-making process. Crisp bread is sometimes called delikatess-bread (Karp *et al.*, 1966; Kuukankorpi, 1957; Roos, 1955).

For the production of crisp bread, rye with low amylase activity is required. The whole meal flour should have a specific particle size distribution curve (Gustafson, 1955):

—about 20% over 1000 μ
—about 30% over 600 μ
—about 50% over 150 μ

The amylogram maximum for crisp bread flour should be at least 400 BU combined with a low maltose value. Crisp bread production is now fully automated. The most important factor for crisp bread bakeries is the uniformity of the raw materials, especially the flour.

PUMPERNICKEL

Pumpernickel (Rotsch and Schulz, 1958), a well-known special rye bread, originated in Westphalia, West Germany. It is produced from 100% rye meal with the sourdough process. The baking time is extremely long, 18–36 hr, starting at a temperature of 150° C and decreasing to 110° C. This bread does not develop a normal crust. Its crumb is dark brown in color (which develops during the long baking period at low temperatures) and the taste can be described as bitter-sweet. It has a very high dextrin content.

Pumpernickel bread has a very long shelf-life. It is now produced according to many different recipes and baking conditions in different countries. If the dark brown color is produced by adding molasses to the dough and not by the baking conditions, the bread is not a true pumpernickel.

V. OTHER USES OF RYE AROUND THE WORLD

Only a small amount of the total world rye production is used in human nutrition. In West Germany, for example, one million tons of rye out of the three-million-ton crop is milled into flour and whole meal for production of bread and other products for human consumption. The development of industrial uses of rye is a major concern of rye-growing countries. Undoubtedly the animal feed industry will remain as the main user of this grain. In all rye-producing countries, more than 50% of the total crop is now used for this purpose.

A. Animal Feed

A number of factors limit the use of rye as an animal feed. Wieringa (1967) isolated a substance from rye kernels that was considered toxic to animals. This substance was identified as a 5-alkylresorcinol. However, there is no toxicological evidence that this substance is harmful to animals. Plant breeders are trying to select lines of rye with lower quantities of resorcinol (Musehold, 1974). Evans et al. (1973) found that spring rye varieties contained less alkyl-resorcinol than winter varieties. The rye grain has a peculiar taste which is not liked by some animals (Musehold, 1973).

In general, it can be stated that the feeding value of rye is higher than that of barley or oats. Feeding tests with swine and cattle have shown that up to 30%, and sometimes 50%, rye can be used in a mixed feed. Horses fed on rye grain showed no ill effects from possible toxic factors (Antoni, 1960).

B. Alcoholic Fermentation

Rye is an important raw material in the alcohol distilling industry. Special whiskey types, e.g., Canadian whiskey, are produced, in part, from rye. Industrial ethyl alcohol is practically all produced by chemical synthesis. However, all beverage and pharmaceutical alcohol must be produced from cereal grain or other natural products. The quality demands on rye grain for the distilling industry are much less stringent than for bread production (Offer et al., 1970).

C. Other Uses

Special physical properties of rye or rye flour or its constituents are basic to the industrial utilization of this grain outside the food and feed industries. Rye flour, because of its high water-binding capacity, is an excellent adhesive. Small amounts of rye flour are used successfully in the glue, match, and plastic industries. An excellent review of the industrial uses of rye and other cereal grains was published by Adams (1973).

LITERATURE CITED

ADAMS, M. F. 1973. Total utilization of wheat. In: Industrial Uses of Cereals. Amer. Ass. Cereal Chem.: St. Paul, Minn.

AMERICAN ASSOCIATION OF CEREAL CHEMISTS. 1969. Approved Methods of the AACC. 9th edition, The Association: St. Paul, Minn.

AMTSBLATT DER EUROPAEISCHEN GEMEINSCHAFT. 1969. Nr. L 100, April 28, 1969, p. 8.

ANTONI, J. 1960. Roggen als Futtermittel. Landbauforschung 10: 69-72.

ARBEITSGEMEINSCHAFT GETREIDE-FORSCHUNG. 1971. Standardmethoden fuer Getreide, Mehl und Brot, 5th edition. Verlag Moritz Schaefer: Detmold.

ARBEITSGEMEINSCHAFT GETREIDE-FORSCHUNG. 1973. Symposium Schimmelverhuetung. Granum-Verlag: Detmold.

BANECKI, H. 1970. Einfluss des Weizen- und Roggeneiweisses auf das Altbackenwerden des Brotes. Bericht 5 Welt-Getreide- und Brotkongress 5: 179-184.

BANECKI, H. 1972. Veraenderungen der waehrend des Altbackenwerdens des Weizen- und Roggenbrotes enzymatisch isolierten Staerke. Getreide Mehl Brot 26: 6-9.

BENDELOW, V. M. 1964. Modified procedure for determination of diastatic activity and α-amylase activity. J. Inst. Brew. 69: 467-472.

BOLLING, H., and DREWS, E. 1970. Pentosanmenge und -beschaffenheit bei Sortenmehlen. Annual Report 1970, Bundesforschungsanstalt fuer Getreideverarbeitung: Detmold, F 25-F 26.

BRABENDER oHG. 1973. Das Schnellamylogram, Anwendungstechnik Blatt Nr. 1203. Brabender oHG: Duisburg, West Germany.

BREYER, D., and HERTEL, W. 1974. Zur Bestimmung von Proteaseaktivitaeten in Weizen und Roggen sowie deren Mahlprodukten mit synthetischen Substraten. Getreide Mehl Brot 28: 13-16.

BRUECKNER, G. 1953. Neue Verfahren ueber die Kennzeichnung der Mehle mit lichtelektrischen Methoden. Getreide Mehl Brot 3: 49-53.

BRUECKNER, G., and SCHOENMANN, I. 1961. Ueber Anteil und Zusammensetzung der einzelnen Kornteile deutscher Roggen. Getreide Mehl Brot 11: 37-41.

BRUEMMER, J.-M., and STEPHAN, H. 1972a. Einfluesse der Mehlqualitaet auf Brotausbeute und Qualitaet bei Roggen- und Weizenmischbrot. Getreide Mehl Brot 26: 10-16.

BRUEMMER, J.-M., and STEPHAN, H. 1972b. Ueber die Zusammensetzung und Wirkungsweise von Frischhaltemitteln bei Brot. Getreide Mehl Brot 26: 289-295.

CANADIAN WHEAT BOARD. 1974. Canadian Grain Handbook, Crop Year 1974-1975. The Board: Winnipeg, Canada.

CASIER, J., DE PAEPE, G., and BRUEMMER, J.-M. 1973. Einfluss der wasserloeslichen Weizen- und Roggenpentosane auf die Backeigenschaften von Weizenmehlen und anderen Rohstoffen. Getreide Mehl Brot 27: 36-44.

CREMER, H.-D., SCHIELE, K., WIRTHS, W. and MENGER, A. 1969. Die ernaehrungs—physiologische Bedeutung des Brotes. Fortschr. Med. 87: 1257-1260.

DOERNER, H., and STEPHAN, H. 1956. Ueber pH-Untersuchungen an Teigen und Broten. Brot Gebaeck 10: 171-176.

DOOSE, O. 1964. Neuzeitliche Herstellung von Roggenvollkorn- und Roggenschrotbrot. Hugo Matthaes Verlag: Stuttgart.

DREWS, E. 1966. Bestimmung des Wasseraufnahmevermoegens bei Roggenmahlprodukten. Muehle 103: 187-188.

DREWS, E. 1968. Beurteilung der Bestimmungs - methoden fuer die Auswuchsschaedigung bei Roggen im Bereich hoeherer Amylaseaktivitaet. Getreide Mehl 19: 1-7.

DREWS, E. 1969. Veraenderungen der viskosen Eigenschaften der Roggenschleimstoffe durch Temperatur, Saeuerung und Salzzusatz. Annual Report 1969, Bundesforschungsanstalt fuer Getreideverarbeitung: Detmold, F70-F72.

DREWS, E. 1970a. Beschaffenheitsmerkmale der Pentosane des Roggenmehles. Brot Gebaeck 24: 41-46.

DREWS, E. 1970b. Studien ueber die Wirkung von Saeure und Salz bei Herstellung von Roggenbrot. Brot Gebaeck 24: 141-145.

DREWS, E. 1971. Quellkurven von Roggenmahlprodukten. Muehle 108: 723-724.

DREWS, E. 1972a. Einfluss der Quellstoffbeschaffenheit auf das Backverhalten des Roggenmehles. Getreide Mehl Brot 26: 154-157.

DREWS, E. 1972b. Untersuchungen ueber das Pufferungs- und das Saeuerungsvermoegen bei Roggenmahlprodukten. Brotindustrie 15: 169-181.

DREWS, E. 1973a. Erfahrungen mit Methoden zur Erfassung des Ausmasses der Staerkeenzymolyse im Hinblick auf die Brotbeschaffenheit. Getreide Mehl Brot 27: 189-197.

DREWS, E. 1973b. Schwankungen der Mehlqualitaet (Type 997) in Abhaengigkeit von der Beschafftenheit des Roggens. Getreide Mehl Brot 27: 305-311.

DREWS, E., and REIMERS, H. 1965. Einige Qualitaetsmerkmale von Roggenpassage-mehlen in Abhaengigkeit vom Diagramm. Muehle 102: 51-53, 67-70, 108-110.

EVANS, L. E., DEDIO, W., and HILL, R. D. 1973. Variability in the alkylresorcinol content of rye grain. Can. J. Plant Sci. 53: 485-488.

GOLENKOV, V. F. 1965. The formation of rye gluten. Biochemistry of grain and of breadmaking, collection 6, pp. 120-125. Academy of Science of the U.S.S.R.: Moscow.

GOLENKOV, V. F., PANKRATIEVA, S. A. and PURGINA, G. F. 1969. Aenderungen im Gehalt der freien Zucker bei Winterroggen waehrend des Keimens. Vestn. Sel'skokhoz. Nauki 3: 42-44.

GORDON, J. 1970. Basic production aspects of rye bread. Baker's Dig. 44: 38-39, 67.

GUSTAFSON, V. 1955. Anforderungen an Schrote fuer die Knaeckebrotherstellung. Brot Gebaeck 9: 156-158.

HAGBERG, S. 1952. Gums in rye and wheat. Report II pp. 339-356. Congrès International des Industries de Fermentations.

HOLAS, J., HAMPL, J., and PRIHODA, J. 1973. Einfluss der Pentosane auf die Qualitaetsmerkmale der Roggenmahl-produkte. Getreide Mehl Brot 27: 273-280.

HUBER, H. 1964. Die Kochsalzwirkung bei der Verarbeitung von Weizen-und Roggenmehl. Brot. Gebaeck 18: 21-28.

INTERNATIONAL ASSOCIATION FOR CEREAL CHEMISTRY. 1960. Practical method for determination of moisture in cereals and cereal products, Standard No. 110. The Association: Vienna.

INTERNATIONAL ASSOCIATION FOR CEREAL CHEMISTRY. 1971a. Determination of "besatz" in rye, Standard No. 103. The Association: Vienna.

INTERNATIONAL ASSOCIATION FOR CEREAL CHEMISTRY. 1971b. Method for determination of α-amylase activity of flour, Standard No. 108. The Association: Vienna.

JANKIEWICZ, M. 1975. The protein complex of bread dough as an interacting system. Die Nahrung 19: 775-783.

KARP, D., KLING, R., and GARPING, B. 1966. Erfahrungen bei der Untersuchung von Roggen fuer Knaeckebrot und andere "Fladenbrote." Brot Gebaeck 20: 169-174.

KUUKANKORPI, P. 1957. Einige Probleme bei der kontinuierlichen Herstellung von saurem Knaeckebrot. Brot Gebaeck 11: 190-194.

LEE, J. W., and RONALDS, J. A. 1972. Glycosidases and glyconases of wheat flour dough. J. Sci. Food Agr. 23:199-205.

LEMMERZAHL, J. 1955. Die Dextrinmethode zur Bestimmung der Auswuchsschaedigung bei Getreidemehlen. Brot. Gebaeck 9: 139-142.

MOETTOENEN, K. 1969. Stationary reaction stage in colorimetric α-amylase assay. J. Sci. Food Agr. 20: 279-286.

MUSEHOLD, J. 1973. Zur quantitativen Bestimmung einer toxischen, phenol-haltigen Substanz des Roggenkornes. Z. Pflanzenzuecht. 69: 102-106.

MUSEHOLD, J. 1974. Grundlagen fuer die Zuechtung eines 5-Alkyl-Resorcinfreien oder -armen Roggens. Dissertation Univ. Hamburg.

NEUKOM, H., GEISSMANN, T., and PAINTER, T. J. 1967. New aspects of the functions and properties of the soluble wheat flour pentosans. Baker's Dig. 41: 52-55.

NEUMANN, M. P., and PELSHENKE, P. F. 1954. Brotgetreide und Brot, 5th Edition. Verlag Paul Parey: Berlin.

OFFER, G., GOSLICH, F., and HALDENWANGER, M. 1970. Die Verarbeitung von Roggen zu Alkohol unter Verwendung von Darrmalz oder technischen Enzympraeparaten. Branntweinwirtschaft 110: 151-154.

OXLEY, T. A. 1948. The scientific principles of grain storage. Northern Pub. Co. Ltd.: Liverpool.

PELSHENKE, P. F. 1941. Die Berliner Kurzsauerfuehrung. Muehlenlaboratorium 11: 105.

PELSHENKE, P. F. 1951. Neuere Ergebnisse und Erfahrungen ueber den Aufbau und die Zusammensetzung des Getreidekornes. Getreide Mehl Brot 1: 45-50.

PREECE, J. A., and MacDOUGALL, M. 1958. Enzymatic degradation of cereal hemicelluloses. II. Pattern of pentosan degradation. J. Inst. Brew. 64: 489-500.

REHFELD, G., and KRAUS, S. 1961. Beitrag zur Bedeutung der Roggen-schleimstoffe fuer die Frischhaltung von Gebaecken. Ernaehrungsforschung 6: 82-95.

ROHRLICH, M. 1960. Phytin als Pufferungssubstanz im Sauerteig. Brot Gebaeck 14: 127-130.

ROHRLICH, M. 1961. The biochemistry of rye bread production. Baker's Dig. 35: 44-50, 70.

ROHRLICH, M., and HERTEL, W. 1966. Untersuchungen ueber den Eiweissabbau im Sauerteig. Brot Gebaeck 20: 109-113.

ROHRLICH, M., and HITZE, W. 1970. Aktivitaet und Verteilung der Cellulase im reifenden Roggen sowie der α-Amylase im Keimling und Endosperm. Getreide Mehl Brot 20: 17-23.

ROOS, A. M. 1955. Die technische Entwicklung der Knaeckebrotfabrikation. Brot Gebaeck 10: 171-174.

ROTSCH, A., and SCHULZ, A. 1958. Taschenbuch fuer die Baeckerei und Dauerbackwarenherstellung. Wissenschaftliche Verlagsgesellschaft m.b.H.: Stuttgart.

ROTSCH, A., and STEPHAN, H. 1966. Qualitaetsverbesserung bei Roggenmischbrot durch Enzyme. Brot Gebaeck 20: 1-4.

RUEBENBAUER, T., and BISKUPSKI, A. 1953. Experimental estimation of the baking value of rye by means of the sedimentation method. Hodowla Roslin, Aklimatyzacja I Nasiennictwo 2: 459-478.

SCHAEFER, W., and FLECHSIG, J. 1973. Das Getreide, 4th edition. Alfred Strothe Verlag: Hannover.

SCHULERUD, A. 1934. Der Saeuregradbegriff. Z. Ges. Getreidew. 21: 29-32, 68-71, 134-136.

SCHULZ, A. 1947. Die Grundlagen der Berliner Kurzsauerfuehrung. Baecker Konditorzeitung 2: 1-3.

SCHULZ, A. 1960. Hinweise fuer die Verhuetung der Schimmelpilzbildung auf Brot. Merkblatt Nr. 40 Arbeitsgemeinschaft Getreideforschung: Detmold.

SCHULZ, A. 1966. Fundamentals of rye bread production. Baker's Dig. 40: 77-80.

SCHULZ, A., and STEPHAN, H. 1958. Untersuchungen ueber die kontinuierliche Sauerfuehrung. Brot Gebaeck 12: 128-129.

SCHULZ, A., and UHLIG, H. 1972. Einsatzmoeglichkeiten von α-Amylasen in der Baeckerei unter besonderer Beruecksichtigung der Bakterien-Amylase. Getreide Mehl Brot 26: 215-221.

SEIBEL, W. 1968. Vorschlaege zur Feststellung des Gesundheitszustandes von Getreide und zur mechanischen Besatzbestimmung. Muehle 105: 471-472.

SEIBEL, W. 1975. Beitrag des Brotes zu einer bedarfsgerechten Ernaehrung. Deut. Lebensm. Rundsch. 71.

SEIBEL, W., and DREWS, E. 1973. Charakterisierung der Qualitaetsklasse "Brotroggen". Muehle Mischfuttertech. 110: 483-484.

SEIBEL, W., DREWS, E., and REIMERS, H. 1971. Sortenabhaengigkeit von Qualitaetsmerkmalen beim Roggen. Untersuchungsergebnisse an in der Bundesrepublik vermehrten auslaendischen Roggensorten. Z. Pflanzenzucht. 66: 103-150.

SPICHER, G. 1970. Voraussetzungen fuer die Vermeidung von Schimmelinfektionen bei Brot und Backwaren. Merkblatt Nr. 65 Arbeitsgemeinschaft Getreideforschung: Detmold.

SPICHER, G., and STEPHAN, H. 1969. Kuehllagerung von Sauerteigen- eine Moeglichkeit zur Rationalisierung der biologischen Teigsaeuerung. Industriebackmeister 17: 9-15.

STAATSVERLAG DER DDR. 1973. DDR Standard Roggen (TGL 2755), Amt fuer Standardisierung der DDR. Staatsverlag der DDR: Berlin.

STEIN, E. vom. 1971. Das neue Monheimer Salz-Sauer-Verfahren. Brot Gebaeck 25: 131-133.

STELLER, W., WIRTHS, W., and SEIBEL, W. 1974. Brot als Lebensmittel soziologisch und technologisch betrachtet. Fortschr. Med. 92: 159-164.

STEPHAN, H. 1960. Detmolder Einstufenfuehrung. Merkblatt Nr. 41 Arbeitsgemeinschaft Getreideforschung: Detmold.

STEPHAN, H. 1963. Einfluss der Kleberqualitaet des Weizenmehles auf die Qualitaet des Weizenmischbrotes. Brot Gebaeck 17: 52-55.

STEPHAN, H. 1970a. Die Aussage von Laborergebnissen ueber den Verarbeitungswert von Mehl und Schrot. Brot Gebaeck 24: 231-239.

STEPHAN, H. 1970b. Zusammenhaenge zwischen Verpackung und Qualitaet des Brotes. Industriebackmeister 18: 92-94.

STEPHAN, H. 1970c. Zweistufige Sauerteigfuehrung mit 2,5 - 3,5-stuendiger Vollsauerreifezeit. Merkblatt Nr. 64 Arbeitsgemeinschaft Getreideforschung: Detmold.

SUOMELA, H., and YLIMAEKI, A. 1970. Die Veraenderungen bei der Lagerung von feuchtem Getreide. Bericht 5. Welt-Getreide- und Brotkongress 3: 29-36.

THOMAS, B. 1964. Die Naehr- und Ballaststoffe der Getreidemehle in ihrer Bedeutung fuer die Brotnahrung.

Wissenschaftliche Verlagsgesellschaft m.b.H.: Stuttgart.

THOMAS, B., and LUCKOW, H. 1969. Ueber den Staerkeabbau waehrend der Teiggare in Abhaengigkeit von der α-Amylase. Brot Gebaeck 23: 24-28.

THOMAS, B., and TUNGER, L. 1965. Die elastischen und plastischen Eigenschaften der Roggenkrume in Abhaengigkeit von der Backzeit. Brot Gebaeck 19: 220-224.

U.S. DEPARTMENT OF AGRICULTURE. 1970. Agricultural Marketing Service, Grain Division. Official Grain Standards of the United States. U.S. Government Printing Office: Washington, D.C.

WASSERMANN, L., and DOERFNER, H.-H. 1974. Raster-Elektronenmikroskopie von Gebaecken. Getreide Mehl Brot 28: 324-328.

WEBERPALS, F. 1950. Fundamentals of rye bread production. Baker's Dig. 24: 84-87, 93.

WEIPERT, D. 1972. Rheologie von Roggenteigen. 1. Mitt. Ueber die Moeglichkeiten der Viskositaets-messungen an Roggenteigen. Brot Gebaeck 26: 181-187.

WIERINGA, G. N. 1967. On the occurrence of growth inhibiting substances in rye. H. Veenmann en Zonen NV: Wageningen, The Netherlands.

WOLTER, K. 1974. Auswertung von Untersuchungsergebnissen gewonnen bei backtechnischen Ueberpruefungen von Mischmehlen. Brotindustrie 17: 50-53.

ZIMMERMANN, R. 1968. Die Helligkeitsbewertung von Mehlen und Backwaren. Baecker Konditor 22: 66-69.

ZIMMERMANN, R. 1969. Die Ergebnisse der Helligkeitstypisierung der Mehle. Agroforum 3: 225-227.

INDEX